WOMAN
IN THE ORIENT

BY

JACOB BABA YOHANNAN, M. A., M. D.

PRESS OF
A. R. FLEMING PRINTING CO
ST. LOUIS.

CONTENTS.

	PAGE.
INTRODUCTION	5

CHAPTER I.
| PERSIA | 9 |

CHAPTER II.
| PERSIAN POETRY | 13 |

CHAPTER III.
| LIFE AND DEATH OF PERSIAN WOMEN | 33 |

CHAPTER IV.
| GIRLHOOD AND MAIDENHOOD | 40 |

CHAPTER V.
| THE WEDDING | 51 |

CHAPTER VI.
| POLYGAMY | 70 |

CHAPTER VII.
| MARRIED LIFE | 79 |

CHAPTER VIII.
| WOMEN'S ATTIRE | 89 |

CHAPTER IX.
| OCCUPATIONS | 99 |

4 CONTENTS.

PAGE.

CHAPTER X.
SOCIAL LIFE.. 110

CHAPTER XI.
RELATIONS OF WIFE AND HUSBAND........................... 123

CHAPTER XII.
SICKNESS AND DEATH................................. 130

CHAPTER XIII.
THE ONLY HOPE... 143

APPEAL TO BAPTISTS.. 159

Introduction.

I WAS born in Oroomiah, Persia, September 25, 1872. Before I was old enough to remember my father, he died, leaving a family of two girls and three boys to the tender care of my mother, who nurtured us with her innermost love and guided us with rare wisdom. At the age of eight I was sent to the little mission school, which I attended during the six winter months for several years. At fifteen I was at the high school. Later on at the mission college, where for several years I was intimately associated with the missionaries, who inspired me with the desire to go to America, where I could have better advantages for study.

On leaving college I went home and told my mother and brothers of my purpose of coming to America, who, hoping to deter me from carrying out my plans, at first refused to defray the expense of my journey; but when they found how earnest was my desire for a better education, they at last consented and gave me the needed aid.

I was nineteen when I left Persia to face the unknown difficulties of the journey on land and ocean, made harder for me on account of my not being able to speak any but my own, the Persian, language.

In New York and Chicago I had great difficulty in making myself understood, for I found no one who could speak with me, and so had to endure hardships that would otherwise have been avoided; but my heart did not fail me, and I felt that God would care for me and guide me aright.

Later, in the following fall, I went to Alton, where I was converted to the Baptist faith and was baptized by Doctor Abbott, and was sent to Shurtleff College for three years, where I studied hard; but I was still greatly hindered by my ignorance of the English language, and also, because I wanted to use all my time for study, I suffered financially. Leaving Shurtleff, I went to Ewing College, where I graduated in 1897.

In the following autumn I went to Rochester Theological Seminary, and afterwards to the Southern Baptist Theological Seminary. My health would not permit me to complete the course, so I went to Texas for six months. Returning, I stopped in St. Louis and was induced to fit myself, through a course of study at the Beaumont Hospital Medical College, for I hoped for work in the Lord's field as a medical missionary.

I am profoundly grateful to the noble and worthy faculty of the Beaumont Hospital Medical College for their personal interest in me, and for the generosity and kind treatment during my connection with the college; especially am I grateful to Doctors Dorsett, McCandless and Auton, for their kindness in allowing me the use of their personal scholarship. I am also grateful to all other friends, and to many Baptist people who have interested themselves in me during the nine years of religious connection with them. May God repay them for their noble deeds, and may I glorify God in my work the better for having had a better understanding through their help!

J. B. YOHANNAN.

PERSIAN SHAH (MEUZAFER-EDIN).

CHAPTER I.

PERSIA.

PERSIA is called Iran by the natives, because it was the birthplace of the Arian race. It extends nine hundred miles west and east, and seven hundred miles north and south. At the present time it is bounded on the north by Asiatic Russia, on the east by Caspian Sea and Afghanistan, on the south by the Persian Gulf, and on the west by Asiatic Turkey. It has an area of 636,000 square miles. The land is ranged by high and fertile mountains. The people can raise most everything they desire. The table-land is irrigated everywhere. The pure water, healthy fruit, and unchangeable climate on the one hand, the roses and flowers of all kinds on the other, make Persia a paradise for mankind.

The ground during the winter is covered with two or three feet of snow, and in summer it is adorned with all that is pleasant to the eye. There is something marvelous about Persia as a country. It has the same laws and the same sort of people as in the time of Cyrus; nothing has changed except religion; while other kingdoms of the world have undergone many changes, Persia retains her old glory.

The history of Persia has furnished many infallible facts to the progress of civilization and Christianity. It was at one time the seat of the patriarch Abraham, prophet Daniel and wise men. It was the kingdom of Persia that protected Christianity at her very infancy. It was the kingdom of Persia that granted religious liberty to the Jews and released them from captivity. The Persians are still of the same disposition towards Christianity, were it not for the despotic and degraded teaching of the Koran.

The present population of Persia is about 10,000,000. In religious belief, 100,000 are Christians, 60,000 Armenians, 30,000 Jews, 15,000 fire worshipers, while the rest of the 10,000,000 are Mohammedans. The principal cities are Teheran, the capital, which has a population of 250,000; and Tobriz, the second capital, 200,000. The present ruler is Mu-zofer-Eden-Shah, who occupies the throne of his father, Naser-Eden-Shah, after his assassination on May 1, 1897.

It is said that he was the best ruler that
Persia ever had since the time of Cyrus.
He ruled fifty years without any trouble
in the form of his government. The present Shah has already proved himself in
regard to some future changes in the
crises of Persian civilization. May God
conform his mind to that end which may
provide an open door for the spread of the
gospel of Christ among the Persians and
Mohammedans.

The government of Persia is an absolute monarchy, so far as it is in harmony
with the principles of religion. If the
Shah does not comply with the religious
teachings, he will be stoned to death.
This is the only reason why he is not able
to give way to civilization.

The Persian rulers have always been
very fond of pleasure. When the Shah
goes out he has from 2000 to 3000 people
following him, and most of them are sup-
ported by him. He must furnish them
with horses, dress and food. The number of women is not limited to the Shah.
He keeps from fifty to one hundred and
fifty wives. The previous Shah had one
hundred and fifty wives, who were presented to him from all parts of the coun-

try; but I am not able to state what became of them after he was assassinated. One would be very much surprised to see so many wives in the harem fighting each other like cats around milk, but when the Shah enters the harem they all keep quiet.

During my eight years' residence in America it has been my privilege to be a member of two literary societies, namely, "Alpha Zeta" of Shurtleff College, and "Logossion," of Ewing College, Illinois, from which I have acquired most of my knowledge regarding the real office of women in the world.

It has also been my good fortune to address a number of ladies' organizations in different States on the subject of "Persian Women," who are, as a rule, without culture, refinement or Christianity.

In this little volume I hope to give the readers some of my personal observations and thoughts in regard to Persian women, which may be of interest to the women of America, who are so freely and fully enjoying their womanhood under the light of civilization and Christianity.

CHAPTER II.

PERSIAN POETRY.

THE Persians are the most romantic
and sentimental people in the Orient.
Their intimate contact with Arabian literature and poetry has beautified the
pages of their history with the passionate
love and burning emotion which are
characteristic of the Arab. The sentiment
of their poetry and pathetic feeling of
their songs cannot be equaled even in the
masterpieces of Milton and Shakespeare.
The passionate inspiration and false im-
agination of their writers have wandered
in the mysterious desert of sorcery, and
have secured ample materials from their
ranging mountains and lofty hills, from
their forest rivers and flowing fountains,
which send their crystal waters to the
beautiful and fragrant roses of the valleys;
and as the roses bloom, they send their
timely gifts to perfume and delight the
hearts of true lovers in the Orient.

The roses with their delightful blos-
soms smile at the rising sun and fill the
atmosphere with fragrant odor. The Per-

sian thinks that roses are the first gifts of God to mankind. Thus says one of the poets at the time of his death:

"Oh, bring to me some of your morning roses,
  And let me die in their fragrant air."

In Persian sentiment the flowers and roses indicate that the dying person is going to paradise.

Woman, indeed, with her charming beauty and attractiveness, makes this planet of ours a paradise for man, "for without the woman the other materials cannot create any inspiration and sentiment in the mind of a writer of poetry," says the poem. The singers have their violins upon their shoulders, and they go about uttering their pathetic stories of love.

PERSIAN POETICAL MYTHOLOGY.

The object of every true writer of poetry, both in the Orient and in modern times, has been the bringing of the obscure history of the past into the light of the present. The voice of present civilization and culture is forever speaking in behalf of the most noted men of the past. These, in fact, are nothing more than witnesses of the providence of God. It is quite true that some must labor while others enjoy the fruits of their labor.

The story of Ishtor, the Queen of Beauty and Love, the Goddess of Romance, is full of pathos. Ishtor was the daughter of Sen, the Moon-God—corresponding to the Greek Venus. She is represented in the legend as dressed in perfect splendor, with her rings and jewels, going to Hades, the region of darkness and shadow, searching for her missing husband, Tommuz. Upon her arrival she found the gates of Hades fastened. She went to the porter and said: "Oh, keeper of the door, open the gate that I may enter in. If thou openest not the gate, I will break it and attack the entrance, and will raise the dead, who will devour the living." The porter was very much alarmed at the indignant utterance of Ishtor, and politely he replied: "Stay, lady; do not break down the gate. I will go and tell this to Queen Nincigal." The porter entered and told Nincigal: "Thy sister Ishtor curses thee." The Queen of Hades trembled on hearing of the indignation of Ishtor, and said to the porter:

"Go, open the gate to her, but shake her like any other common woman."

Ishtor went to the first door of Hades and she was admitted, but her crown was taken away from her head; at the second door her ring was taken, at the third door her jewels, and finally she was stripped of her last garment. At each gate she exclaimed: "Porter, do not take away my jewels and garments;" but the porter paid no attention. All these things she endured because of her ardent love for her missing husband. After she had thus diligently searched Hades, there was very much interest in her behalf among the gods. They wished to rescue Ishtor from the shadow of hell. As the result of her mission, all of the gates were thrown open before her, and she got back all of her jewels which had been taken away, and resumed her honorable dignity as the great imperial Goddess of Love and Beauty.

LITERATURE.

If Greece was proud of Homer, Italy of Dante, and England of Shakespeare and Tennyson, Persia is equally proud of her Omar-Khayam, Hafiz Nizami, and Sodey, whose zeal and inspiration have been a stimulus to thousands of young people. One of the most impressive poems is that of Nizami of Ganja, who lived in the twelfth century. He added more love songs to Persian literature than any other poet of his time. This is one of his productions, entitled "Lailie and Majlem:"

Lailie was the name of a young lady, and Majlem was that of a young man. Lailie was the daughter of a poor Arab, yet very handsome indeed, and Majlem was the son of a Chieftain. These young people were deeply in love with each other. Their love was of such devotion and intensity that when their songs are sung by the people of the Orient (Persia), their sentiment arouses sympathy among all classes of people. Once when Majlem had fixed his anxious eyes upon Lailie's beautiful face, he sang:

"The soft expression of her face,
  Brought distraction to my burning brain;
  No rest I found by day and night,
  She was forever in my sight."

As Lailie's people were in the desert, they one day moved their tent and went to the mountains with their family, leav-

ing no trace of their march behind, and
making every possible effort in order that
the two young lovers might not com-
municate with each other. Majlem became almost insane in searching after his
sweetheart through wilderness and mountains. Finally his father became alarmed,
and, taking with him a band of strong
friends, according to the custom, went
hunting the people of Lailie, and found
them in the mountain. He made his
proposition to them concerning the marriage of his son Majlem to their daughter
Lailie. But his proposal was unfavor-
ably answered. Then the indignant and
disappointed old man returned homeward.

Poor Majlem, with a broken heart, fell
on the floor; there he sat and wept bitterly. One day, while walking about the
camp of the Arab, he was seen by some
relatives of Lailie. She recognized him
when her attention was called to him, but
she dared not go out to meet him on account of her unsympathetic father, lest he
would become angry. But anxiously
from morning to evening she gazed about
in the hope that she might get her dear-
est Majlem once more. While sitting
beside the fountain, under the shade of a

willow tree, near the encampment, with
an anxious and broken heart she sang:

"Oh faithful friend and true lover,
Still distant from Lailie's view,
Still absent beyond my power,
Go bring thee to my fragrant bower;
Oh noble youth still thou art mine,
And Lailie still is thine."

While thus singing her love songs
under the cool shade of the willow tree, a
stranger by the name of Ibie-Solem passed
by. His eyes gazed upon her attractive
face; he frankly went to her father, and
entreated him for his daughter's hand.
Fortunately, his visit was favored by the
old man, because he appeared in a kingly
dress, and consequently they were united
in a quiet, simple way. Then poor Majlem attempted to induce his friends to
fight the unjust Arab, but he could not
succeed in his plan.

After the marriage the new husband
brought his costly gifts with a long line
of camels, burdened with robes, beautiful
rugs and carpets, silks of all kinds, and
the most valuable things, to be laid at the
feet of his worthy bride.

The strains of all kinds of music for
marching, which announces the coming

of the bridegroom dressed in the richest
cashmere and smiling at each step,
were heard.   However, all these things
availed nothing to make Lailie's married
life happy, because her unwise father did
not allow her to marry the one whom she
loved more than all that Ibie-Solem had.
It was simply done by the act of her cold-
blooded father.   She became the wife of
Ibie-Solem, but she still cherished the
memory of Majlem.

"Deep in her heart a thousand woes
  Disturbed her days' and nights' repose;
  A serpent at its core writhing and gnawing
    evermore,
  And no relief but a prison room,
  Being now at the lonely sufferer's door."

The rolling years did not bring any
cure to the aching heart of Lailie.   She
sat peacefully in her prison-tower, watching the course of the sun by day and
flashing of the stars by night, but with a
never-ceasing longing in her heart for
Majlem.

One day while she was sitting in her
chamber and meditating on her fate, she
heard a noise of wailing.   There was great
confusion in the family.   A messenger
came in with a note, announcing the sud-

den death of her husband, Ibie-Solem. Although the news was a star of hope to her heart, yet, after the manner of the Arabs, she put on the garment of sorrow and wept with the rest. When the years of widowhood were fulfilled, she freed herself from the harem, and calling her trusted servant she sent a happy message to Majlem, her old sweetheart. She appointed a certain time and place to meet him. She made her way through the growth of palm trees and roses, and did not stop until she saw the familiar face of her cherished lover. Stepping gently to his side and laying her hands around his neck, she said: "Oh, Majlem, have you come?" As he recognized the familiar voice and gentle touch, he was overcome with emotion, and fainted at her feet.

"His head, which in the dust was laid,
Upon her lap she drew, and dried
His tears; with tender hands she pressed
Him closely to her breast;
Be here thy home beloved, adored,
Revive the blest, oh Lailie's Lord.

"At last he breathed, around he gazed,
As from her arms his head he raised;
'Art thou,' he faintly said, 'a friend,
Who takes me to her gentle breast;
Dost thou in truth so faintly bend
Thine eyes upon a wretch distressed?'

" 'Are these thy unveiled cheeks I see,
  Can bliss be yet in store for me;
  Is this thy hand, so fair and soft,
  Is this in sooth my Lailie's brow?'

" 'In sleep these transports I may share,
  But when I wake it is all despair;
  Let me gaze on thee, even tho' it be
  An empty shade alone I see;
  How shall I bear what once I bore,
  When thou shall vanish as before?' "

To this Lailie replied readily:

" 'Here in this desert join our hands,
    Our souls were joined long before;
  And if our fate such doom demands,
    Together wander evermore.

" 'Oh, Majlem, never let us part,
    What is the world to thee and me?
  My universe is where thou art,
    And is Lailie all to thee.' "

Majlem knowing that according to the law of the Arabs he could not make her his wife, with a low voice he answered:

" 'How well, how fatally, love,
  My madness and my misery prove;
  All earthly hopes I could resign,
  My life itself, to call thee mine.
  But shall I make thy name spotless,
  That sacred spell, a word of shame?

" 'Shall selfish Majlem's heart be blest,
And Lailie prove the Arab's jest?
The city's gates though we may close,
We cannot still our conscience's agony.
No, we have met, a moment's bliss
Has dawned upon my gloom in vain;
Life yields no more a joy like this,
And all to come can be but pain.' ' '

Then he pressed her close to his aching
heart and kissed her for his last good-bye.
Accompanied by her servant she went back
to her home and lived a lonely life.   The
time of her life's sunset drew nigh.   She
called her mother to her bedside and entreated that when she was dead, Majlem
might be allowed to weep over her grave.
After she was gone the faithful servant
carried the sad tidings of her death to the
broken-hearted Majlem.   He made his way
sorrowfully to her grave and wept bitterly
about a week.   At last he was found resting his head on her grave, and died, like
his beloved, of a broken heart.   Then her
grave was uncovered, according to his request, and Majlem was buried beside her.

"One promise bound their faithful hearts,
One bed of cold, cold earth united them when
    dead;
Severed in life, how cruel was their doom,
Never to be jointed, but in the silent tomb."

Another work of Nizami is his story of Shirin and Parhod, whose love affair is no less impressive than that of Lailie and Majlem. In poetical discourses the Persian admires this story as one of the best productions. Shirin was the espoused of King Khosroe Pareviz, and Parhod was a famous engineer in his employ. This couple fell in love with each other, and consequently the king became aware of the fact. However, the king promised to give her up to Parhod if he could execute the plan of bringing into the city all the water that is in the mountains. Parhod sat himself to the herculean labor and to the demand of the king. He had almost completed the work when the king heard that Parhod had almost finished his course of labor. Fearing he was surely going to lose his espoused Shirin, according to his own statement, he then sent a messenger to Parhod, who was hard at work for the sake of his beloved Shirin, telling him that she was dead. When Parhod got the sad news he became filled with despair and threw himself down from the lofty top of the mountain and killed himself.

"Do not make a jealous promise,
To thy fellow-man, when thou
Art not right in thine heart."

LOVE SONGS.

We have a great number of unwritten love songs in Persia. The musicians hang their instruments, such as violin and certain other kinds, over their shoulders, to sing and play. They usually go in companies of two; while the one is singing the other will give his attention, and when he is through the other will sing. Now, if one cannot understand what the other sang or meant, he has to give him all the money they take in and, also, his own instrument. One thing is very interesting in comparison with what the musicians do in the civilized world, and that is this: they have no written songs or music of any kind, and they must memorize all their songs. They usually learn so many songs, one would naturally be surprised to hear them singing for so long a time different songs, all from memory. All popular songs in Persia are stories of love, with the exception of a few, which are religious.

ASLEY AND KARAM.

These lovers lived several centuries ago in Oroomiah, Persia. Asley was the daughter of a Nestarian, a man of much money and influence. The young lady had great reputation for charm and gracefulness. She spent most of her time around the fountains in her father's beautiful garden, caring for flowers, vines and other plants. She dressed always in flowery apparel, embroidered with silk of great value. The decorated gardens, the handsome young girl, and the freshly flowing fountains by the wayside were the special objects of attraction to the passing pilgrims.

Karam was the son of a wealthy Mohammedan, who lived in the neighboring village. One day while he was hunting, his "lala-man" (servant) was holding in his hand a "kurgoon," an oriental trained hunting bird, which is kept for that purpose; and he let her fly into a group of birds that the other birds might rise to his gun. The bird made a peculiar sound, attracting the attention of the master and servant to the living picture at the fountain. Karam, beholding the beauty and grace of her countenance, loved her from

that time, and made up his mind to marry
her.

From that day he gave up his hunting
pleasures, and sat down under the shade
of a palm tree. Inspired by her beauty,
he wrote poems upon the palm leaves and
sent them to her by his "lala-man" (servant). Asley, in reply, wrote poems of
love to him about the place of woods and
forest, because that was the place where
they fell in love with each other. She
sent her message of love by his "lala-
man," who told the glad news to his master, Karam.

As the years passed the lovers grew
impatient. Finally Karam besought his
father to call upon her father and request
that she might become the wife of his son,
Karam. But there seemed to be great
difficulties in the way, because Asley was
a Christian and Karam a Mohammedan,
and there is no association of any kind
between Christians and Mohammedans,
either socially or religiously. Finally
the bishop of the Christians gave his religious decision to her father that it would
be absolutely impracticable and impossible for him to give his daughter to a Moslem. Karam's father went back disap-

pointed.   But Karam kept on going to
the woods, a place near to Asley's home,
and sang his songs of love and wept over
his disappointment for ten years, so that
her father could not persuade her to marry
any one else.   But, of course, the ten
years' period could not cool their burning
love for each other.   Her father sold
everything he had and took his daughter
and fled to Russia, in order that he might
get rid of the young Moslem.   But the
love-sick Karam told his father good-bye,
and followed the track of his sweetheart
to Russia; and, while traveling, sang his
love songs all the way, thus:

"Oh, justice, I appeal in behalf of my misfort-
    une.
  I have left my friends and fatherland,
  I am a grief-stricken wanderer,
  After my missing black-eyed Asley.
  Though far thou mayest sojourn,
  It will not discourage me from following;
  If necessary, I will enter into thy church,
  And bow with confession before thy cross."

Asley and her parents settled in a
Russian village.   One evening while
Asley was sitting on the house-top, looking over the beautiful mountains, she
heard a voice singing the strains of a

Persian song. She recognized Karam's voice, and the strange land seemed to her as though it were the place where she first met him, and she was filled with joy. After singing some of the songs that were very familiar to her, he thought he had better change his appearance, so that he might go into her house without being recognized by her parents, and thus get an opportunity to see his loved one. He let his hair and beard grow long and put on the dress of a Dervish, a sect of Mohammedan religious men, and went to the door and knocked, just in the same manner as the Dervishes do in Persia. The servant came to the door. He said to the servant: "I am a man of God; have suffered from toothache for several days, and desire the help and kindness of your mistress to give me relief." The mother of Asley had some experience in curing such pain as that of toothache. The servant reported the man's trouble to her. She opened the door for him and welcomed him to her house, and promised him relief. She called her daughter Asley to hold his head in her lap while she would pull out his tooth. Both the daughter and the patient wept with bitterness. The mother thought the Dervish wept because of his severe pain, and her daughter because she was too sympathetic. After the first tooth was pulled, the Dervish said: "Pull the next out; it pains me very much." She pulled that one, too, and then he asked her to pull the next, until every one of his teeth were pulled out. After having no more reason for staying longer, he removed his head from Asley's lap and said: "I have had thirty-two teeth pulled out, but did not feel any pain because my head was in the arms of my beloved." After singing a song he entreated that he might be permitted to rest for the night at her comfortable home. She granted his humble request. On the morrow he made himself known to her parents. Finally they consented to their marriage. But the end of their devotion was rather pitiful. After the wedding it grew very cold. Asley drew her lover's seat near the fire. The natural fire from without and the love fire from within developed into a visible flame. Asley was frightened to see her dear husband burning to death; hastily she took a pitcher of oil, thinking it to be water, and poured it upon him; but unfortunately it only increased the flames, which consumed both of them, and in a short time they were reduced to ashes. Such was the true devotion of faithful and loyal lovers, and such was the unhappy end of their career.

PASSION SONG BY HAFIZ.

"Sweet maid, thou wouldst charm my sight,
  And bid these arms thy neek enfold;
  That rosy cheek, that lily hand,
  Would give thy poet more delight than
  All Bokhara's haunted gold—than all
  Gems of Samarcand.

"Boy, let your liquid ruby flow,
  And bid thy pensive heart be glad;
  Whatever the frowning zealots say,
  Tell them their Eden cannot show a
  Stream so clear as Roenobad—
  A bower so sweet as Masellog.

"Oh, when these fair perfidious maids,
  Whose eyes our secret haunts infest,
  Their dear destructive charms display,
  Each glance my tender heart invades
  And robs my wounded soul of rest,
  As tarers seize their destined prey.

"In vain with love our bosoms glow;
  Can all our tears, can all our sighs,
  New lustre to those charms impart?
  Can cheeks where living roses blow,
  Where nature spreads her richest dyes,
  Require the borrowed gloss of art?

"Speak not of fate; oh, change thy theme,
  And talk of odors, talk of wine,
  Talk of the flowers that round us bloom;
  It is all a cloud, it is all a dream,
  To love and joy thy thoughts confine,
  Nor hope to pierce the sacred gloom.

"Beauty has such resistless power
  That even the chaste Egyptian dame
  Sighed for the blooming Hebrew boy;
  For her how fatal was the hour
  When to the banks of Nilus came
  A youth so lovely and so coy!

"But oh, sweet maid, my counsel hear;
  Youth should attend when those odes
  Whom long experience renders sage,
  Which music charms the ravished ear,
  While sparkling caps delight our eyes,
  Be gay, and scorn the frowns of sage.

"What cruel answer have I heard?
  And still yet by heaven I love thee still!
  Can aught be cruel from thy lips?
  Yet say, how fell that bitter word
  From lips which streams of sweetness fill,
  Which naught but drops of honey sip?

"Go boldly forth, simple lay,
  Whose accents flow with artless ease;
  Like orient pearls at random strung,
  Thy notes are sweet, the damsels say;
  But ah, far sweeter if they plan
  The nymph for whom these notes are sung!"

CHAPTER III.

LIFE AND DEATH OF PERSIAN WOMEN.

OUR readers might conclude, from what has been said in the preceding chapter, that the Persian women are held in the highest esteem and admiration, both by their husbands and by society. On the contrary, in Persia, as well as in all the Orient, woman is expected to minister to the passions of men. In fact, their societies are made up according to their traditional and religious sentiment. Tradition is the infallible teaching of their ancestors, while their religion is so-called inspired teaching, so strongly and emphatically dictated in the Koran. Their religion asserts the absolute inferiority of women. The light of the present age has been very little reflected into the social, mental and spiritual life of women in Persia.

We will now discuss the conditions in which the Persian women live and die, as they are religiously observed by all.

THE BABYHOOD OF THE WOMAN.

This period covers about two years, during which the child should be nour-

ished.. The birth of the child is full of
agonizing anxiety to the mother, as the
physician is never called in such cases.
The patient is absolutely under the care
of some ignorant female attendants, who
know nothing about this important branch
of medical science. After the child is
born, they first take care to distinguish
the sex. If it is a boy, the servants and
friends have a joyful time in breaking the
glad tidings to the father and his friends,
from whom they usually receive gifts of
all kinds. Then hearty congratulations
will be extended by neighbors and relatives to the parents of the child; more especially to its mother.

After three days have passed, the father
will consider it his great duty and privilege to invite to his house all those who
have known of the happy event, and en-
tertain them, providing music and all
manner of festivity in order to make it a
memorable occasion, because to him a son
is given, one who will perpetuate his name
and protect the home of his family. But
if it is a girl, the servants, who are anxiously awaiting the news, will feel very
disappointed, because there will be no
gifts given, no music, and no festivity.

ARMENIAN WOMAN BESIDE THE CRADLE.

and happy, and may you be blest with many sons and no daughters.'' In fact, the birth of a girl is regarded a misfortune to her parents. It is a traditional saying that if a man has four daughters, he is on the road to poverty; and if he has seven, he is a curse to the community. The reason for this is that the daughter is lost when she gets married; she brings no benefit to her parents, except to leave them in debt: but the son cares for them in their old age.

The child is usually salted a week after its birth, according to ancient custom, and thus made subject to all traditions. If the child is boy, the mother is not allowed to show him to the public until he is two or three months old, fearing that by looking at him they might cause his death. A charm will soon be brought, upon which the prayer of Mohammed is written, and will be fastened on his right arm, in order that he may be free from the ''evil eye,'' or other sickness. No one is allowed to use a flattering expression, such as ''What a handsome boy he is!'' lest the ''evil eye'' might cause him to die. The mother will keep him as comfortable as possible, and when he cries she is always ready to rock him to sleep by her motherly songs.   But this is not the case with a baby girl.   At the latter part of the first week of her birth she will be put in a hard wooden cradle and wrapped tight in old swaddling clothes. She will be left therein most of the day and all night.   When she cries the mother will not pay the slightest attention to her. The only thing she does is to rock the child, without any other motherly care. The child will keep on crying and crying, until it is utterly exhausted. The father will have nothing to do with it at all.   It usually takes several months before the cold-hearted father will even smile at his child.   If the little thing is fortunately pretty, she might win her way to his affection; but if she lacks such attractiveness, he will not take her into his arms till she becomes useful in housework. This is especially so with a young married man who has for the first time become a father.   The little ones usually receive a better treatment among the wealthy people.   They always have nurses and servants for the children.   The mother is so dignified and proud that she will have nothing to do with her children, except in case of absolute necessity. The mother is not as such recognized by the children. They grow up under the care of nurses and servants; and when they are old enough to be married, they take their nurses and servants with them to their new homes.

The cradle is made of a hard kind of wood, and affords no comfort of any kind to the tender life of the infant. The child is generally laid in the cradle, flat on her back, and is fastened tight with a bandage to the upper extremity of the cradle. But before they do this they will fasten the arms of the child to its sides and its feet straight out. Then they will wrap the body in some old and heavy bedding, without taking off the clothes of the child. They tie these materials so tightly that the poor little thing cannot move at all. But that is not all: they veil her head and eyes also during the sleeping hours. This will be the condition of the child during the first two years of her life.

The mother, as a rule, is so ignorant that she has no idea of the constant predisposition of the infant towards diseases of all kinds; nor does she care for its proper food, which is so essential for pres-

ervation of the child's life. She will herself eat any kind of food, and give it to
her little ones without knowing the harm-
ful effects that may follow; and when
either the child or herself gets sick, she
thinks it is according to God's plan, and
not the effect of careless diet.

NAMING OF THE GIRL.

The boy is given a name on the day of
circumcision by the priest who circum-
cises the child. But there is no certain
time for naming a girl. On some day,
however, an old gray-haired woman
will be invited to the house, who will,
perhaps, take the ten days' old girl on
her arms and cry into the infant's ear by
calling any name that is suggested to her
by the parents of the infant. She says:
"This is your name."

Most of the names of women have a
meaning in Persia—such as "Almas,"
diamond; "Sherin," sweet, etc. These
names are very popular. They have no
family name: by this I mean they don't
use any. If there is some confusion of
names, they will distinguish them from
each other thus: "Almas, daughter of
Smith;""Sherin, daughrer of David," etc.

CHAPTER IV.

GIRLHOOD AND MAIDENHOOD.

IN THE preceding chapter I described
the life of a girl, extended from birth
till two years of age. Now I will de-
scribe the child-life. I have been very
much impressed by the hearty attention
and treatment that the people in America
render to the children in general, and
especially to girls. When I am in Sun-
day-school, or at an entertainment or
picnic, I enjoy seeing the people manifest
such kind consideration for the happiness
of the children. I remember, in the State
of Illinois, I was watching a young
woman who was teaching her infant class
the Christmas songs in such a kind and
gentle manner that I cannot forget it.
She was going around with a smiling face
and clapping hands, not only that the
children might learn the songs of their
Savior, but also that they might enjoy
the inspiration of that hour. After she
got through she dismissed them, but
there were two sweet little girls whom she
kissed, just as a mother does, and then I

asked: "Are these little girls your daughters?" She smilingly replied: "Oh, no;" and she said: "What made you think that?" I said: "Because you kissed them so tenderly." She said: "I always do that when I see them." I said nothing more, but thought of many little girls in Persia who are without Christmas songs and such teachers who can sing and play with them in the name of Christ.

There is nothing more pleasant to Dr. W. W. Boyd, the pastor of the Second Baptist Church, St. Louis, than to see children sing and play. He spends an hour every week teaching them spiritual songs. With a Persian priest it would be quite a ridiculous thing to spend an hour in conversation with children, and more so with Persian women. When a person becomes unselfish and childlike in order to please and amuse the children, the people would think he was losing his common sense. The trouble with the Persians is that the words of Christ are strange to them: "Suffer the little children to come unto me and forbid them not, for of such is the kingdom of God." The Savior was so much interested in the life of a child because he himself had been a child, and he appreciated the importance of tender care. This period usually begins at three and ends at seven years in Persia, which is very short in comparison with that in America. During this period there is nothing done for the mental and spiritual welfare of the children by their parents and relatives. They are treated like cats—eat their meals and stay at the house because they are girls, and from that time must learn to confine themselves to domestic affairs. They are rarely taken to places of amusement as they are in America. The parents will attend all kinds of entertainments and festivities and leave their little ones at home, because they are ashamed of being parents of girls.

Of course, the little girls do not mind these things, as they are not yet old enough to take an interest in social life, consequently will not worry about them so long as they have not seen them. But there is one thing that every child naturally knows, and that is a love of play and amusement. This love of play has in many cases become a twofold factor of physical and intellectual development in

the life of a child.   It is a natural disposition of the human race that some men can
never gain intellectually, physically and
morally unless they come in touch with
the force and spirit of education, while
others possess that from their childhood,
as a result of their earliest and independ-
ent development.   And so a love of play
in the early life of a child creates an am-
bition to become great in the destiny of
life, no matter whether it is a boy or girl.

In order that their childhood may develop they, of course, must be provided
with a place to play, which is very much
needed among the common classes.   During the summer they usually get along
nicely by going in the street to play; but
in winter they have a very hard time.
The common people in general have one
large room, where the entire family spends
the winter, and in that room they will do
their spinning, grinding and cooking.
There will be no room left for the children to play at all.   The little folks must
then get up upon the house-tops, which
have been kept clean and dry, because the
snow falls so deep that they cannot get
out in the streets during the winter.   One
would be surprised to see men, women

and children all spending their leisure time upon house-tops in cold days. Now, that is the only place for the children to enjoy themselves; but sometimes they are not allowed even there, because they think the children are running around so much as to bother those who are within the house, and consequently the parents will strictly keep them in the house and make them behave as though they were old maids.

The little girls have no artificial games, such as dolls, horses, and all such toys as one sees in this country, which so much encourage the children's development and furnish amusement for them. The girls in Persia do not grow so rapidly as they do in America, because of the lack of happiness and freedom in their lives. Whenever the girls want to have a doll to play with, there is not anything to be bought for them. Finally they will try to make them by themselves from wooden materials, which seldom resemble the artificial doll.

Another fact about the life of a girl in Persia is the lack of a proper training in the manner of living. While the Persians are very particular about the code

CHILDREN WITHIN THE HOUSE.

of living, the little girls get no benefit from it. The mother considers it waste of time to teach her little daughter the manners and habits which they should have. She thinks that they will get better when they become of age. One would be surprised to see the naturally pretty girls coming to table with dirty hands and faces, who could be made like the royal children simply by washing their hands and faces. In Persia, as in every other country, the children grow up after the manner of their parents; and parents are, to a great extent, responsible for the bad habits of their children. A Mohammedan will always be glad to see his boy using vulgar language, stealing and lying, because he thinks his boy is pretty smart; and often the mother and father instruct him to steal.

Another thing they very much need is religious training. This is most essential for a child. The Moslems never attempt to teach their children the religious requirements till they are fifteen years of age. The reason for that is that the children, who have not sufficient understanding, are not responsible for their acts, and consequently need no religious instruc-

tiou. During the fifteen years of their ir-
responsible actions they, of course, build
a fine residence in their hearts for the
devil; and when it is well finished and
furnished, it generally appears pretty
hard for the devil to leave such a palace
of youth.

In my judgment the thing the little
girls in Persia most need is a Christian
mother, who will feel it her first duty to
tell them about Jesus and his wonderful
love for them. A Christian mother can
accomplish much more than a preacher,
with all of his pulpit power. There is no
other influence in the world that would
and could instill the Christian law and
principles into the heart of a child as that
of her mother. Her love alone can lead
the little ones into the path of all happiness and highest good that pertains to the
life of a child.

THE MAIDEN LIFE OF A WOMAN.

This period begins at the age of seven
and ends at sixteen or seventeen, by which
time every girl is supposed to have been
well qualified for marriage. This period
is better appreciated by the mother: first,
because they are helping the mother in

housework; second, because they are active in preparing themselves for their
future lives.   Every mother at this time
will exert herself to the utmost to further
the accomplishments and attainments of
her daughter, because she wants her to
get married as soon as possible. The parents are usually very proud when their
daughter is in great demand, even if she
is ten years old; but if she has no one to
marry her, or if she becomes an old maid,
it would be twice as distasteful to the
mother as it would be to the daughter herself.   Therefore, the parents at this time
will exercise a great deal of care and influence over their daughter. She must
confine herself to the house and be quiet
in her manner; walking freely in the
streets or with young men is strictly
prohibited. She goes out and manifests
good behavior.. Her occupation up to this
time has been only housework, such as
washing the dishes, cooking, and preparing "kyloon," a Turkish smoking pipe.

The Persian girls, as well as women,
are great home-lovers and domestic in
their tastes. They are taught to do
almost anything before they get married.
The principal occupation for the girls

NOBLEMAN'S DAUGHTER IN HER STUDY.

ence to the Koran than a great many Christians do to the Bible. If they, with their dead hope, expect their children, as well as themselves, to read the Koran daily, which is but a false witness of eternal life, how much more Christian people and parents should expect their children to read and reverence the Bible of Jesus Christ, the only true witness and revelation of God's infinite love for man-kind.

It seems to me that there is nothing in the world that is so essential to the life of a woman as education. My personal observation of woman's capacities leads me to believe that women can accomplish wonderful things in the progress of the world, when their faculties are properly developed. The young women of America, as well as of Europe, demonstrate this by filling well the highest positions in every department of life, especially when their education is a Christian education: with strong moral principles in their lives they are a great power in modern society. Christianity not only elevates women's lives to the highest possible degree, but gives them favor in the eyes of their employers and in the community where they live.

The Persian women are of similar nature and faculties, but they are not developed. They can generally use good judgment, but imperfectly. I believe they could have as good schools as there are anywhere else, but religious despotism and traditions are depriving and robbing them of these precious opportunities. So far as I can judge, there is no hope for Persian women to live womanly lives and get rid of these religious superstitions and traditions until they have the freedom of the Gospel of Grace: "If the Son therefore shall make you free, ye shall be free indeed." This means all that the Persian women need.

CHAPTER V.

THE WEDDING.

GOD, in his first expression regarding woman's place and happiness in the world, said: "It is not good for man to be alone." So we see that it is a divine determination and order with which every man and woman must comply, or else their happiness will be destroyed. We also know that man is a social being and must live in society, and in this way comes in contact with different individuals. He knows what a higher and lower life means among the human family. Through such knowledge the young couple are able to originate and develop their love for each other. But degraded customs and corrupt religious principles have led the Persians to believe that love has nothing to do with matrimony. Therefore it is the duty of parents to act as agents for their sons and daughters in selecting a suitable companion, without any reference to their like or dislike of each other. Of course, such marriages increase the possibility of divorce.

Marriage is of practical importance. Whenever a young man or woman neglets it at the proper age, according to the custom of Persia, they will be blamed and cursed; and if they are not able financially, they must borrow money. A good many young men have no money of their own, and it would take them many years before they could save enough to marry. As a matter of fact, the parents are somewhat compelled to provide for them.

The men never live a bachelor life, because Mohammed has given them permission to get married at any time and to have as many wives as they may desire. The young couple have very little to do with the choice of each other. They cannot write, and so they are unable to correspond with each other. Custom does not allow them to go with each other and have familiar conversation in regard to their future life. One often marries without having any personal knowledge concerning the girl whom he is to marry, except what he has been told by his people.

If the parents or relatives of a young lady should see her going around with a young man, as the young people do in the civilized world, they would surely kill her. The only way is for the young people to see each other at a distance and make love in that way, and often they talk to each other by a sign of the eye with a correct understanding. If the girl smiles at the young man while he is acting in this manner, he thinks she loves him or cares for him. By these means they develop a certain kind of courtship. In using the eye-sign they mean business and never deceive each other.

THE BETROTHAL.

The usual time of engaging a young lady to a young man is between seven and fifteen years, although a girl is betrothed as soon as she is born. This is a matter of promise between the parents of both parties when they are special friends. They will make a contract and wait till the little couple grow up. Sometimes this contract is made before their birth. They make a promise something like this: "If I have a daughter and you have a son, I will give her to him for wife; but if you have a daughter and I have a son, I would like to have her for my son." They will sign the agreement and that is all. This plan may be sug-

gested in a social manner when they are
in a social gathering, and when they
have made such an oath in the presence
of a social assembly they cannot go back
on it, even if it costs them their lives.
Just as Herod could not go back on his
rash oath, although it cost him the be-
heading of John the Baptist.

THE BETROTHAL CEREMONY.

First of all, the parents of the young
man will generally make the selection of
a certain young lady with the consent of
all the intimate relatives. After that
they will send a private message to the
young lady's father, as to whether or not
he will accept their proposition in regard
to the matter. In reply to the message
he will request of the father of the young
man time in which to consider the stated
proposition with the different members of
his family and perhaps near relatives. A
week later the father of the young man
will repeat the same message and ask for
the final reply, in answer to which he
will either receive an acceptance of his
proposition, or a rejection. Now, supposing that his offer is favorably received.
Then he will send two of the prominent

men of the town to the home of the
young lady, in order to get a personal and
definite answer. Having secured this,
they will return with favorable news to
the young man's house, and get ready
for a betrothal entertainment. Then the
young man's father will extend invitations to his relatives and friends, whom
he takes along with him, together with a
beautiful ring, a pair of shoes, a head-
dress and some money, perhaps from five
to six dollars. In addition to this, he
provides a great deal of candy, meat, rice
and other necessary things for a respectable festivity, sending them about two or
three days ahead, so that they will be
useful at the time of the entertainment.

Now, at the time of the entertainment,
the young man's father with his respect-
ive men and women, will arrive at the
town and will be met by the people of the
young lady and welcomed to their homes
as well as to her home. Of course, the
guests usually take one meal at her home,
and at other times will be entertained by
the town people. However, after supper,
which is the customary time for such an
occasion, the father of the young man will
mention the object of the meeting in a

brief address, in which he will indicate his gratitude, first to the people of the young lady, and then to his own men for their personal interest in the occasion. He will next be followed by the father of the young lady, who will also make a brief and happy speech in response to the frivolous address, and to his relations, in which he will say something like this: "This girl does not belong to me but to her grandfather or grandmother." They will transfer the same expression to somebody else, until all in turn have the honor. They usually begin from the oldest to the youngest in the family, and then to the intimate friends. Of course, this means nothing to her father, except as a matter of respect and good-will towards his family and friends. In this way they will give their hearty consent to the proposed betrothal ceremony. The father then gives his real and definite answer, in which he publicly says to the father of the young man: "I give my daughter to be a pair of shoes to your son;" that is to say, "I am perfectly willing to do so." Then it becomes the duty of the young man's father to get up and take his hat off and kiss the right hand of every relative of the girl, to express his gratitude for the great honor.

Meanwhile, the ring will be carried to the ladies' apartment, where the girl and her friends are at the time. It will be handed to her by two elderly women, ''gray-haired,'' who will address the girl by saying: ''Your father and all your relatives have taken the liberty to betroth you to Mr. John Smith. If you are willing to obey him, take this ring and put it on your finger.'' She takes the ring and puts it on her finger, as she is expected to do. The old women who took the ring will come back before the audience and testify to it; and if she did not take the ring, they must testify to the same. But usually the girl takes the ring and does what her father desires. If the girl in any way refuses to take the ring, no one has any right to compel her to do so; then, of course, the engagement is broken, which, as a rule, gives a very unpleasant reputation to both parties.

The length of the engagement varies from six months to three years, but the usual time is six months. Sometimes they marry so young they can hardly tell the difference between a betrothal and marriage. During that time the espoused

girl must not show her face to the friends
and relatives of her sweetheart, but keep
in a veil. On certain days the young
man will call and see her, but not very
often, because the public will talk about
both parties. Sometimes he is not allowed
to see her at all. I think it is a pretty
hard thing for a groom to be away for six
months and not see his sweetheart. That
would not suit the American youths at
all; and yet they don't know what it is to
live in a Christian land, and grow up
under Christian culture, where their love
for each other is naturally developed!

During her engagement the young
woman wears her outdoor costume, in
which she is completely hidden from her
lover; but she has the advantage of seeing
him through the small openings of her
facial veil, and engraves on her heart his
beloved figure as he comes and goes
through the street. If he meets her on
the doorway or street, face to face, he may
or may not be able to derive benefit from
the meeting, for often he does not recognize her.

I remember there was a man who had
two daughters. The one was very beautiful, while the other had a pretty face,

but only one eye. One day a young man went to that man's house with the intention of making love to one of his daughters, as is the custom for young men to do in Persia. Of course, they all knew the young man's purpose, as well as his date of coming. They were well prepared to welcome him. Then they invited the better looking girl to come forward, so that he might see her appearance. She came, and was liked at the first sight. The young man went back to his home and told his story of love in a respectable manner to his people, and they made arrangements to get the girl with an expensive betrothal. After the betrothal period was over, which is usually about six months, then came the time of marriage. During this time he had spent a good deal of money for dresses and entertainments, in order to have his bride at his home, and she came. When the poor fellow saw her face he was terribly disappointed, because they sent him the girl with one eye instead of the good-looking one. Of course, after spending so much money it was too late to pursue another course. He had to be resigned. The cause of such deceitful disappointment is the strictness of their blind custom, which does not allow a free conversation to the candidates.   The young people of America must not forget to value the freedom which they enjoy concerning matrimony at this most enjoyable period of life.

THE DUTY OF THE YOUNG MAN'S PARENTS TOWARD THE ESPOUSED GIRL.

It is a common thing among the Oriental people during the betrothal period on certain days for the parents of the young man to send some appropriate gifts to the new members of their family. This is done according to the different seasons.   If it is summer, they will take a quantity of fruit; if it is winter, they will take a pair of shoes or a dress, as the occasion requires.   The girl's parents do not ask for anything.   This is only a matter of respect and close relationship between the two families. But negligence in the matter of these little gifts will always bring disappointment to the girl's parents and displeasure to the community.

THE WEDDING GARMENTS.

Extravagance is one of the special features of a Persian wedding. From the time of their engagement to each other to the time of the wedding the parents of both parties must spend a large sum of money for garments and entertainment purposes. Of course, the young man's people usually spend more than the young girl's. The wealthy class can stand the pressure all right, but the poor, who compose most of the population of the country, become paralyzed by the burden of debt, under which they suffer for lifetime.

Several weeks before the wedding takes place, both parties send their representatives to buy the wedding trousseau, and they spend several days in selecting and purchasing it. They usually buy several different garments for the bride, of which one suit must be a Persian shawl, while the others may be common goods. But for the bridegroom they will order only one suit. It may cost very much or very little. Still, he is satisfied, because he knows they will not pay as much attention to him as they do to her. All of the money expended must come out from the pocket of the young man's father. After they are through purchasing, the parcels are taken to the bride's home. Then she extends an invitation to her young lady friends to come and help her in cutting and sewing the wedding garments, so that they may be ready for the wedding.

THE WEDDING INVITATIONS.

They are altogether different than those of America, such as sending cards by mail, and most of them will not perhaps be received till the wedding is over. But in Persia several young men will be sent out, two by two, to the villages and towns where they have acquaintances and relatives (Matt. xxii. 3; Mark vi. 7). They use this expression as an invitation: "Shlama Allakoon;" meaning "peace be upon you," or "come upon you." This is the way Christians salute each other, but the Mohammedans say: "Salam Alakoom," which means the same thing. Then they extend their invitations by saying: "Mr. David has a marriage for his son. He says this happy event does not belong to me, but to you. Come to the marriage with all of your family; everything is ready." The young lady's parents will extend similar invitations.

The messengers are expected to be very sober, polite and faithful, so as not to forget or neglect any family, and not spend their time on the road by talking to any one, but be devoted to their mission. If there is any official in any way connected with the families of the couple, the parents should go in person to the officer and take along with them a sheep or a piece of sugar, which they present to him, asking him to attend the wedding. Then in return he will express his best wishes for the bridegroom, and if he intends to attend the wedding he will bring his gift either to the bride or to the groom.

In case of an ill-feeling between the relatives and the parents of the couple, the invitations must be extended to the relatives either by the parents themselves or another person of influence, in order to pacify and persuade them to attend the wedding; and if they still insist upon not coming, the groom must himself go to them and plead with them (Matt. xxii. 1-14); and if they still have no regard for the groom's invitation, they are left alone.

THE WEDDING.

This ceremony is quite another thing from that of America. I remember when I was a student at Shurtleff, only a few

months after coming to America, one day
there was a wedding to take place at the
M. E. Church in the afternoon. Up to
that time I knew nothing about the American marriage ceremony. However, I had
a friend who took me along with him, and
we went to the wedding, but when we got
in, the church was crowded. We took our
seats in the back pews, and then I asked
my friend: "Where are the wedding
parties?" He said: "They will be here
directly." Then we saw them marching
to the altar. When I saw the bride and
groom with their arms linked together and
the flowers in the bride's right hand kept
almost in contact with nostrils of the
groom and in front of his face, I became
surprisingly excited, and could not help
laughing. Finally I got out of the
door and laughed until I became exhausted, and then came in again. Of
course, I caused the people to laugh also,
because they knew that I was a stranger to
such a scene. Anyhow, I did not see any
veil on the face of the bride, any music
or dancing. I thought that was a very
poor wedding occasion indeed. Well,
then we saw them standing in front of
the altar and the minister questioning

them, and they answered quite loud, and finally they were dismissed with a benediction. The special feature of the occasion was my laughing, for they all looked at me instead of looking at the bride.

In Persia the wedding occasion usually lasts from two to seven days, with several hundred invited guests, with dancing and music of all kinds, which, as a rule, excite the merry hearts.

At the wedding feast the invited guests, as well as others who are present, are given a place at the table alike. From five to fifteen sheep are killed, or from four to six oxen (Matt. 22: 4). A large quantity of rice and other necessary things are provided. Three meals a day continue from the beginning to the end of the marriage. They sit on the floor around the long table, and eat with their fingers, according to the custom of the country. But each individual must occupy his seat according to his rank and garment, whether he is a stranger or a relative. If he dares to occupy a seat of which he is not worthy, when somebody else comes the new-comer will have his seat, and the offender will be asked to take a lower seat, because he did not have on a wedding garment (Matt. 22: 11, 12). - After the meal they usually have dancing. If it is in the day time, they go out-of-doors and engage in dancing and singing. The place for these is an open courtyard. The women and men, hand in hand, form large circles; the one at the leading end will hold a handkerchief in his hand, swinging it in the air and waving it slowly around the circle; the musicians, walking in the center of the circle playing the flute and drum, make an exciting noise. Gradually they become enthusiastic, leaping frantically till they are exhausted; then they drop out and some one else takes their place.

At night they have the dancing within the house, and they usually arrange a large house for that purpose. At the time of retirement it is the duty of the neighbors to take the guests who come from abroad and care for them over night and bring them back in the morning for breakfast. Of course, they eat their breakfast where they are, but it is the duty of the groom to keep up his three meals a day and early in the morning bring the guests from the different homes to his own. In order to do this the musicians will play about fifteen minutes early in the morning, signifying the calling back of the guests. Often the musicians and several others will go after them in order to get them to come.

On the second day they prepare to go after the bride, if she is living at a distant town. Some hundreds of young men go, some on foot and some on horseback. Arriving at the bride's home they are entertained for several hours, and sometimes are kept over night. Next morning the bride is dressed in her wedding apparel, over which is thrown a large red veil, covering her entire person. Bidding her people farewell, she is mounted on horseback and kept in position by two strange young men, who must either be related to her or to the groom, till they arrive at the house of the groom. Here a large crowd of people extend to her a hearty welcome, emphasized by fire-works, dancing and music. Each of the young men who have followed her are presented with one or two chickens by her people, which are called "bridegroom birds." This manner of wedding is practiced by the people who live in the villages and towns. But in the cities the procession is usually made at night so they can have torches and make the occasion a pleasant event (Matt. 25, 6-12). Both in towns and cities, where the procession is within a short distance from the groom's house, two young men will be sent ahead to break the news to the groom and his people that the bride is coming. Then they dress the groom, and with his friends follow to meet the bride. Having arrived there, he and his party bow their heads before the bride, and then he throws an apple over her head; thus he joins her in marching homeward. On their arrival they are greeted by shouting crowds. Then the groom and his party of young men go on the house-top, where they watch the music and dancing, while the bride is taken to the doors of the prominent citizens and relatives before going to her father-in-law's home; as a matter of respect, however, and not of compulsion. At each door she will receive gifts from the family, showing their appreciation of the respect rendered by the bride and her people. The gifts are voluntary, and they may give anything they prefer. Sometimes they take a quantity of candy and raisins and throw them before her, and sometimes copper and silver money, as a sign of prosperity. Sometimes they tie a piece of Persian shawl about the horse's neck.

Afterwards she is brought back to the home of the groom, and the entertainment is prolonged for two or more days.

The last day will be devoted to the welcoming of the bride's parents and relatives, who bring her goods and trunk. At a certain hour of the day they open the trunk and let everybody see her trousseau, that they may witness the proper care and preparation of her father. The goods that are found in the trunk are of many different kinds and styles and prices, to fit the worthiness of the guests and relatives, because they are to be given as presents to all who are invited to the wedding at the time of their departure for their homes. On the morrow the guests will all have expressed their appreciation of the entertainment and gifts which they have received, and offered their best wishes for the future happiness of the bride and bridegroom. The couple, instead of taking a wedding trip, stay at home and are quiet.

CHAPTER VI.

POLYGAMY.

PERSIA has been the seat of polygamy from the time of Mohammed, who first of all satisfied his desires with it and then taught it to be of divine origin. Since that time the country has come under the curse of God. In its well is the water of jealousy, deceit, degradation, and murder. Under its influence the true sexual love has been altered to a bestial passion. It has been an agent of heart-breaking, destroying the happiness of the individual, society, and the nation. Polygamy was not originated by Mohammed, as we know that the patriarch Abraham practiced it. But Mohammed practiced and taught it as a factor of progress, otherwise he would not have, perhaps, succeeded. It is said that "polygamy is the flower of pleasure on the pages of the Koran." This was the only means by which his followers attracted many millions to the perfume of degradation and suffocated the divine aspirations of humanity.

In the religion of Mohammed there are two modes of matrimony. The first is called "Addah," covenant; the second is called "Seyah," contract. In the first, the wife becomes an heir to the property of her husband, and he cannot marry another under the law of "Addah." In the second, the women are made servants to the first "Addah," and one may marry as many as he can support without any objection, but he cannot promise them any money. A man never makes his affairs secret when he marries a new wife, no matter if he marries one every month, because it is upheld and observed by the priests.

THE CONTRACT OF MARRIAGE.

In the ceremony of "Addah," or covenant marriage, the application must be legalized about ten days before marriage by a Moollah (priest) in the presence of two men and two women, together with the parents of the lovers. The Moollah will write the application, which is called "Kabin" (license), in which the amount of dowry to the wife is fully stated. The Moollah will be one of the witnesses to the agreement of both parties. Then he

reads a passage from the Koran and offers
his benediction. He goes into the apartment of the bride and asks her consent,
which she grants. Having done this, the
Moollah writes two letters of contract,
one to the bride and one to the groom.
They must exchange letters, which answer
their real contract, so that there can be no
misunderstanding and mistreatment in
case of divorce.

If the husband does not like her after
living together for some period of time,
and wishes to be divorced, he must, in
the presence of those witnesses, together
with the Moollah, pay the full stated
amount of dowry, whatever it may be, to
his wife before he can get a divorce. The
amount is from fifty to one thousand dollars, either in property or in money. Of
course, this makes it usually impossible
for them to marry more than one woman
under the law of "Addah." The conditions under which a man may marry
under the law of "Addah" are: First, she
must be pure, or a virgin; second, she
must be fruitful; third, she must be free
from any kind of disease. If she lacks
either of these conditions, she will be
divorced without being paid the stated
amount of dowry as a punishment.

MARRIAGE UNDER THE LAW OF "SEYAH."

According to this manner of marriage, a man may have as many wives as he can take care of, and does not have to promise them any dowry. One may make a contract with a woman for certain months or weeks, and when the time is up let her go; or one may make a contract with a woman down-town where he is working and claim her as his wife so long as he is working in that neighborhood, and come back home to his other wives at night. A man who gets a wife under this law of "Seyah" must have a letter of agreement with her, written by a Moollah, in which he states the time limit and the qualities of the wife; as whether religious, married before, or faithful. Under this law he is obliged to support and dress his wives, but not necessarily to promise them any money unless he wants to. They cannot get a divorce until the time is expired. After that they can renew their contract and live together for a longer period, if they wish.

DIVORCE.

A man can get rid of his wife at almost any time, if he complies with the requirements which are written in his contract, 'without any other reason. If he is asked why he wants to get a divorce he will reply, because he is getting tired of her. But she cannot get a divorce whenever she wishes to, because the law does not acknowledge the right of the woman, even if she has a good reason. There is one way by which a woman under the law of "Addah" may secure a divorce, and that is, she may abondon the amount of dowry or money that has been promised to her. Then she will say thus to the Moollah or her husband: "Kabinen, Halal Jonem Azad" ("I abandon my dowry in order that I may free myself"). She will then be divorced without being given any compensation. Some intelligent women manage better—that is, they make themselves disagreeable to their husbands, and provoke them so that they send them away or give them divorce by paying their compensation; this, however, is not often accomplished. But the women who marry under the law of "Seyah" have no way of getting a divorce until their term expires, while their husbands may get a divorce at any time for even a trivial reason. The women and the men of this class are more

like animals than human beings. The
law of divorce is somewhat complicated.
There must be at least two witnesses. It
cannot be done by force or by sending a
message, except in the case of sickness
and dumbness. The husband must use
this expression in pronouncing a divorce:
"Hazzie Talikoon" ("Thou art divorced,"
or "This person is divorced"). This expression is in the language of the Koran.
The wife must have lived separately from
her husband for a month, in order to
make the case effective. If he has but one
wife, he does not necessarily call her name
at the time of divorce; but if he has
many he must clearly distinguish the
name of the one to be divorced.

The evil effects of such laws and marriages upon the real happiness of the
family and society are very grave indeed.
In the first place, the teaching of Islam
has absolutely perverted the original law
of marriage. We find the original law of
matrimony in Genesis: "Male and female
created he them." Being two different
individuals, yet divinely united in one
union, "they shall be one flesh." This
being the fundamental divine principle of
the real state of marriage, we are certain

that the teaching of the Koran has abandoned the true idea of matrimony by giving license for the practice of polygamy, and permission for divorce.

Some might say that this law is justified by the fact that it increases the population; to this I reply that God does not justify such doctrine as this, under the light of the new dispensation, because the law of Christianity alone recognizes the real office of true marriage and secures love and happiness in human society.

In the second place, polygamy is inhuman. What would you think of a mother who gives or sells her ten-year-old daughter to marriage, who does not even know how to put on her dress. Sometimes the innocent child is taken miles and miles away from her mother's home and becomes the wife of an elderly man. In America the girls are very happy, indeed, when they find a chance to marry and be escorted to their bridal carriage or train, but in Persia they cry when that time comes, because of the rough matrimonial road which they must face in a slavish manner. The untimeliness of such marriages is not even recognized by parents in the Orient.

In the third place, it destroys the physical and intellectual capacities of the young girls. They become mothers at an age when they should put their faculties into intellectual development, and their bodies into the health of physical growth. They become impotent in strength and ignorant of domestic duties. They cannot understand *what* and *how many* are their *home responsibilities* in the true sense of home life.

In the fourth place, it removes from them their womanly dignity. When religion and the law claim that a woman is nothing but a slave to the passions of man, then certainly there is nothing left by which she can keep her dignity and virtue.

In the fifth place, it deprives the women of their rights. They become subject to their husbands in all things; whereas they have similar faculties for culture and understanding with their husbands. Whether they know better or not, they must be directed and guided by their husbands like beasts of burden.

Lastly, it corrupts the home and society. It puts children against their mothers, and wives against their husbands, because there is no love in such marriages; eighty-five per cent. of Mohammedans marry without a single element of love. One often observes the several wives, who eat, sleep and live in one house, and their children as well as themselves, fighting each other for the sake of bread, water and lodging where they can be more comfortable. Whenever the husband pays attention to the one who is graceful and beautiful, and also to her children, the others will rush against her and beat her to death. The home is degraded with a continual jealousy, hatred and strife, instead of continual peace and love and happiness.

CHAPTER VII.

THE MARRIED LIFE OF THE WOMEN.

DURING this period the life of the young married woman is different in every respect from her former life. She passes from one sphere of life into another. She can no longer associate herself with the old environments and friends. She must re-form her habits and character. Although young in age, she must be old in manners, because she comes into a circle of two-fold responsibilities. She has to acknowledge her relation to the home as a leader of domestic affairs, as well as to her husband, her best lord, in order to insure happiness. Not only that, but she must realize that she becomes a factor of good example and moral influence among the society and community into which she comes. But it is justifiable to say that in Persia seventy-five per cent. of the women lack the qualifications which may in fact elevate them to the throne of their womanly dignity. They adhere to such things as have been in practice for years in the

past.   This is all due to the influence of
religion, superstition, and the customs of
the country.

THE DWELLING-PLACE.

In Persia the people, poor and rich,
prefer to live together in one community.
They live in villages, towns and cities.
There are no country or farm dwellings,
for the simple reason that they do not
own the farms, which belong to the government; and, also, they are afraid of
robbers, who would burn their houses and
rob them.

The cities have high walls around them,
with as many gates as necessary.   These
gates are closed at eight o'clock at night
and opened early in the morning by the
police guards.   The purpose of these
gates and walls has been to render pro-
tection in time of war, or the sudden rush
made by those of other countries.   There
are many Kurds who live near by who
never work, but live by robbery and
thieving.

Walls and gates are found sometimes
in villages, in order to protect the women.
The men go out to work, perhaps several
miles away, leaving the women and chil-

dren at home. In such cases highway-
men often come to these villages and
towns to take advantage of the good-
looking women and rob them of their best.
If there is no wall around such a town or
village, each family must have a wall
around their own house, with a solid gate
in it, so that women can lock the gate
during the absence of their husbands.
The houses are built in such a manner as
to insure safety and privacy by all means.
In the Orient the people are not so public in their customs as they are in America; whatever is said in a certain family
must be kept strictly secret.

The construction of the houses is of
two classes. The first class is composed
of brick and stone, which occupy a large
lot with a wall around it, and a large, solid
court gate, kept shut day and night. In
the coming and going of the visitors the
guards will open and close the gate. Besides this gate, of course they have a good
many other doors for different parts of the
building. Neighbors are not allowed to
build their houses any higher than each
other; if they do, they are not permitted
to open windows of any size, which might
make it possible for them to look upon

the wives of each other from the window.

In entering the houses of this class of people you will come to the men's apartment first. Here is the head of the family during the day, giving directions to his servants and receiving calls on business. He comes to his private apartment at sunrise, after having a cup of tea or coffee and offered prayer. At dinner time he goes to the dining-room for dinner, and returning spends the rest of the day in his room. The second apartment of the house is called "harem" (forbidden place). This place is for the women and children. No other person can go in but the husband, who comes home in the evening. Generally they are not very particular about Christians going into the harem, because they are pretty sure there would be no danger; but they never admit a Mohammedan, because the Koran very strongly pictures its evil results. The women never veil their face from a Christian, while they must from one of their own faith. The rooms are large and the walls thick, in order to sustain the heavy roofs. The roofs are always flat, made by extending long, large beams across from one end to the other, about

two feet apart from each other. Over
these they put planks and cheap matting
and then the earthy materials, such as
clay and mud mixed with straw chaff.
The roofs become dry and hard, through
which the rain cannot pass, but runs
down in pipes. The purpose of the flat
roofs is to be able to sleep there during
the summer months and to walk and play
during the winter, because the snow falls
very deep, and it is impossible for women
and children to walk, visit and play in the
courts or streets.

The windows face towards the gardens,
which are beautifully planted in the courtyard with choice flowers and fruit trees.
The harem is very elegantly and hand-
somely decorated within. The walls and
the ceiling are plastered with white plaster,
but they have no wall paper. The floor
is finely carpeted with the best kind of
Persian rugs, all of which are manufactured at Shiraz and Hammadan. The
rugs that are sold in America are not of
the best kind, and do not come from Persia, but India and Turkey, sold in the
name of Persia. The Persian rugs are
heavy and thick and soft, with a rich un-
fading color, and have a shining surface.

There are no chairs or tables in the rooms with very few exceptions, for they prefer to sit on the floor, where they can move in any direction they please. Sometimes they have cushions over the rugs. Their garments are made for this reason short and comfortable. They are not very particular about having separate rooms. They rather have one room for eating, sleeping and sitting. They are usually supplied with one room, because they don't have bedsteads that occupy much space in the room. The bedding is all rolled against the wall, or upon a platform made for the purpose during the day, and at bedtime they are spread over the floor again, each finding his or her place to sleep. They have very comfortable and soft pillows along against the walls all round the room, so that they can recline on them and rest.

Around the walls, about four feet above the floor, a row of shelves is made, upon which they put bottles of many different kinds and colors, filled with attar of roses and perfume. Next to their bottles are watermelons and apples, which they keep during winter. The room becomes nothing but a place of art and beauty, where the heart can be merry.

The residences and courtyards, which are arranged in such an artistic manner, where streams of water pass from pool to pool, filled with songs of birds and atmos· phere of roses, in the Persian sentiment, are nothing but "behisht" (paradise). But they are absolutely ignorant that such "behisht," or paradise, is but of very short duration and imperfect compared with the real and perfect and everlasting "behisht," paradise of God.

Next are the houses of common people. This class are the working people and farmers. Some of them have an ordinary living, while others always run behind. These live usually in villages. Their houses are built of mud. They let the water run into the dry earth to be absorbed. Then men and animals will walk upon it until it becomes sticky and compact. The builder takes little by little from the mud, which is handed to him by his assistant, and puts it on the ground. When he completes one layer all around, he lets it dry for a week, so that he can build another layer, and so on. It usually takes him about six months before he can finish one house. When he is through building the walls, then he covers the roof in the same way as that of city houses. But these houses in villages do not have windows like those of the cities. They have one door and two or three small openings in the center of the roof in order to let light into the house. The door is so low that when a person tries to get in he must bow his head. They have no separate apartment for sleeping and eating purposes, but all are done in the one room. Whenever a strange person enters the house, he does not have to knock at the door, as they rarely do that in villages. The lady of the family has no other room where she can arrange her toilet. She runs quickly to a corner of the same room and attends to her toilet: of course, she may be seen just the same.

The floor is not wholly carpeted, but only part of it where people sit; they use cheap carpets and matting; while in another part is a place for cooking, which they call "Tander" (oven). This is circular, about three feet in depth and two in diameter. It has a flue leading to the bottom through which the air passes. Over this oven is a wooden table extending about six feet square, which they call "Kurse," and over this table a blanket is

thrown, which keeps the warm air from
escaping. The women and children will
get under this table all day, and sometimes all night, during the winter. They
enjoy it very much, because they don't
know of anything better. They have no
stoves for cooking and warming purposes,
but the oven is the only place for all
these things. Every day the women
build the fire within the oven, and wait
till it gets hot. The house will be filled
with smoke for two hours until it gradually makes its escape out of the small
openings in the roof and door. The men
will leave the house at that time, and let
the poor women stand the effects of suf-
focation, because the women must attend
to the cooking. This is one of the rea-
sons why the people in the Orient are so
many of them blind and have sore eyes.

The furniture is very cheap and simple.
Only some bottles and earthen vessels on
the shelves with many rolls of bedding
piled on one side of the room. At night,
when spread, it covers the floor; this being the only room in which to sleep. The
houses have a damp and dirty floor; especially when it is raining through the
openings in the roof. Next to this room

is another little room for storing winter provisions, such as melons, grapes, and raisins. A little further off is the stable and barn filled with hay and straw for cattle. The Kurds who live in the mountains have only one large room for sleeping; cows and chickens are in one part; while men, women, and children are in the other. Such is the condition of poor people in Persia.

CHAPTER VIII.

The Women's Attire.

THE women confine themselves strictly to the custom of the country, and are always satisfied with it whenever it is a new one. (They have the indoor and outdoor costumes. The outdoor costume must be used when the women go on the street or visiting. It consists of a "chader," which is a black sheet of cotton or silk, about three yards one way and two the other—something like that of Catholic sisters, covering the figure from head to foot. The veil is of white cotton, tied around the head about the black "chader," covering the face and hanging a little way down in front. It has in it, right opposite the eyes, many small openings, so they can see their way and also look occasionally at the men; but men can neither see their face nor recognize them.)

(The "Jarabs" are a pair of loose trousers of the same material, attached to the others, the appearance of which is quite funny, indeed.)

.The shoes are similar to sandals, made of sheepskin, of red, yellow and blue colors. They cover only half of the feet, and are not attached at the heel, so that in stepping the heels of the sandals clatter in a musical way. They can slip their shoes off and on without touching them with the hands. When they go out in a large group they all look alike, just as a group of birds of the same nature and color at a distance resemble a flock of black sheep in the pasture. No man can tell which is his wife and which is his sister, because if he dares to express his opinion, they all may be his wives or his sisters.

.This costume sometimes is quite inconvenient and uncomfortable, especially in winter; but they have to wear it just the same. Without it they cannot possibly appear in a social circle. Of course, it offers some benefits to the poor women by covering all their cheap and common dresses under it, and yet it looks as if all were new every time. It lasts, perhaps, for a lifetime. If they are old and lack attractiveness, with it they look very much like the well-dressed and attractive ones.

MOHAMMEDAN WOMAN IN HER STREET COSTUME.

The law always protects any lady who goes out in the public places with her costume on. In Persian sentiment there is no worse disgrace than to accost a lady on the street. If she makes public such an insult by any man, in a short while a great crowd will take that man and drag him in the streets like a dog until he dies.

The royal women must always be guarded by a large number of men servants, who go before and behind their carriages; the forerunners marching through the streets with silver rods in their hands, commanding the people to get out of the way. This is not done because they fear anybody, but as a sign of royalty.

The indoor costume is peculiar. This is composed of several short skirts extending down to the knee. They wear from five to ten at one time, one over the other; standing up, they look like a ballet-girl. The wealthy women have their skirts of different colors and trimmings, which consist of borderings several inches deep of real pearls. The trimmings of their gowns are made of silk, satin and velvet, inwrought with gold and silver thread in the most elaborate patterns. This is the indoor costume for summer. In winter, over the skirt and shirt, a waistcoat is worn. ꙮ It is of a popular and fashionable material, with its front about ten inches apart, in order to show the beautiful chemise, and has long sleeves, which are buttoned at the wrist.

The most interesting part of the indoor costume is the head-dress. First of all, they put on a cap made of velvet or shawl material, embroidered at its borders with money hanging to the front of the head. Over the cap is a large triangular handkerchief, tied under the chin, covering the ears, while another part of it covers the mouth up to the center of the nose. This they call "Yashmak." It signifies the noble character and behavior of the women, just as a bridle is used for the conduct of the horse. This "Yashmak" is the common, yet very popular, custom among all classes, poor or rich, old or young, but more especially adopted by the young married women to show respect for their husbands. They need not wear it in the presence of their husbands, but only when other people are present, no matter whether they are the relatives of their husbands or not. The young mar-

ried women must not talk out loud for
the first two years of their married life,
although some of them have been in the
habit of not talking loud for a lifetime.
They must not talk to their fathers or
mothers even in low tones, and the relatives
of their husbands and often strangers are
included, too, for at least several years,
wearing the "Yashmak" over their mouth,
and keeping quiet and respectful. This is
their first duty when they are at their own
home. Sometimes, in case of necessity,
they must talk to the above persons; then
they use children as the mediums, to
whom they say what they want to be
done, and the children take the verbal
message to any of the above persons,
whether father or mother-in-law; and if
there are no children, they talk slowly
and in a low voice without removing their
mouth-cap.

They think this is the best way to
keep peace in the family, because they
usually live together after they get married for several years, no matter how
many sons there are in the family. By
keeping their "Yashmak" over their
mouth they become quiet and peaceful,
otherwise they could not live together for

a year. But this law cannot keep the women from talking even with their "Yashmak" on, because they talk softly, so that the family cannot understand their complaints. The women in the Orient, as well as those in America, have a very active brain for all kinds of talk, and if they were without that mouth-cover they would all become politicians. No wonder St. Paul says: "Let the women keep silence in the church."

When they go to church in a group with mouth-veils on, both old and young, they cannot help but talk and talk about what has been taking place in the community since their last meeting. One would naturally hear the low talk coming from all directions, but could not tell who it is on account of their mouth-veils. I believe if St. Paul knew who it is that talks he would "call them down," but the poor man could not specify the individual. Especially when no man is around, they make a very loud noise. Before one gets through her talk the other begins, so that they may each get the earliest possible chance. If there is any complaint against the women folks in Persia, it is because they talk too much. The husbands are afraid they might cause trouble in the community by their foolish babbling. Persia is not the only country where men complain of women in this respect, but also in America I have very often heard men complain of the same thing. I am quite sure some of them would be very glad if they could adopt "Yashmak" for their wives and keep them in silence, but individual freedom of speech makes it impossible. However, I do not believe in keeping an intelligent woman silent. Any such imposition is only possible in the case of ignorant women. No one would raise an objection to the intelligent and timely suggestion of his wife unless he himself was a boor.

The Persian women, as well as the men, do not wear shoes while in the house, but take them off when they come in, keeping on their stockings if they have any; and if not, they would rather walk in the house with bare feet. Both sexes pay a great deal of attention to the covering of their heads, but care very little for their feet.

PHYSIOGNOMY AND ORNAMENTS.

The Persian women naturally are very handsome and attractive. Their complexion is very bright and clear, their form regular, and their eyes almost always large and black. Blue eyes are not at all a sign of beauty in the Orient, although we have some of them, with very little attention paid them; in fact, girls with blue eyes and light hair are not very much in demand. Since they have no demand, we call them "old maids," even though they be quite young. The real features of beauty are a round face, rosy cheeks, black eyes, black hair, and regular figure. It is true the Persian women have not the education that would enable them to appear in social or educational circles, as those in America, but they will give attractive smiles with their thin lips and white teeth. There is no woman in all the country with artificial or gold-filled teeth. The women, as well as the men, prefer to be fleshy, but not extraordinarily so.

They use powders and drugs for the face, but not often, except where occasion demands it. They dye their hair with a dye which is composed of indigo and

MOHAMMEDAN WOMAN IN HER INDOOR COSTUME.

and enjoy perfect beauty, but lack of money makes it impossible. Such has been the history of humanity: some must bear the burdens, while others enjoy rest.

NESTORIAN WOMAN SPINNING.

CHAPTER IX.

THE OCCUPATIONS OF WOMEN.

THESE are notably different from those of men. The occupations are classified so that men may not interfere with the work of women; and if they do under any circumstances, they are called by a feminine name, which is a mark of disgrace in the Oriental sentiment. If the women attempt to interfere with men's occupation, they will be called also by a masculine name, which is equally disgraceful.

One can observe that this is absolutely contrary to that of America, where the women receive the respect and admiration of men for the work which they are doing, both physically and intellectually. Such freedom cannot be exercised by women in Persia. While it is true that both sexes have equally a divine right to do what may be necessary for their support, yet I disapprove of the enslavement of either men or women to the extremes of fanaticism.

In all cities and towns there is not a single woman employed in store, shop, school-room, or any other business, where in a good many ways they might do a better work than men. When at work, whether at home or abroad, the nature of their work is hard and dirty. Nothing has impressed me more during my stay in America than to see thousands of women engaged in work of all kinds; especially the kind treatment they receive in general at the hands of the employers in any department, and more especially when they are on their way home in the cars, with comfortable seats, while tired men stand. In Persia when women and men are grouped together, instead of men giving their seats to women, the women give their seats to men, because they are regarded as slaves. If there is anything for which God blesses the American people, it is for their respect to the helpless women of all classes.

The occupation of the wealthy classes is limited to a simple work, either intellectual or manual. They superintend the general housework while the servants perform it; even their children are cared for by the nurses. The only thing they do is to eat, smoke their water-pipe three or four times a day, and go to sleep. Oc-

NESTORIAN WOMEN GRINDING.

casionally they embroider and sew just for
exercise.   A very few can read a little,
while most of them do not know how to
spell their own names.   When they have
nothing to ⸱attend to, they will paint
their cheeks and powder their foreheads,
or talk about getting married.   There is
nothing said about the future punishment
through which they must inevitably pass
some day.

The women doing hard manual work are
like beasts of burden.   In addition to the
domestic duties, such as cooking, clean·
ing, and other such things, which they
must do early in the morning, they also
must attend to field work.   Their time is
fully occupied from sunrise to sunset.

First is grinding at the mill.   Almost
every woman must know this. They grind
wheat into flour.   In some parts of the
country where water can be. had, they
have water mills, but where water cannot
be had, the mills are put to action by
donkies and mules; while in other places
women grind at home in the old-fashioned
way.   "Two women shall be grinding at
the mill, the one shall be taken and the
other left" (Matt. 24: 41).   This consists
of two heavy round stones with a wooden

pin in the center fastened into the lower stone, passing through the hole in the upper.   On one side of the upper stone is a fixed wooden handle.   Two women will sit around the mill opposite each other; then they take the wheat and drop it by the handful into the hole at the center of the upper stone, falling between the two stones, when the women turn the mill by the fixed wooden handle.   A large sheeting is placed under the mill, so that in running the flour may fall on the sheet. The flour which is thus made is not good, but under the circumstances they must live on it.

It is interesting to notice the way in which the Persian women bake bread. They first make dough in a large tub and put some yeast in it; for an hour it is covered.   During this time the fire is burning within the oven—the fire-place. Then two women will get ready for the baking by having before them a wooden table and a rolling-pin.   The dough has been cut into small pieces, scattered on a sheeting over the floor.   Then one of the women, usually the younger, takes one of these small pieces of dough and puts it upon the table and flattens it by rolling

PERSIAN WOMEN BAKING BREAD.

the pin over it; then she takes it up and hands it to the other, who strikes it against her arms and hands until it becomes very long and thin and broad; then she puts it on the cushion which is made for this purpose from small twigs of willow trees. She then takes and bakes it in the oven on its hot side, and takes another in the same way until she is through. It takes but a few minutes until they are cooked very nicely. But one thing peculiar about the baker is this: if she has long arms, she makes long bread; but if she has short arms, as a rule she makes short bread. The young ladies who have long arms are always in demand, because they will make long bread when they get married.

HOW THEY CARRY WATER.

The work is altogether done by young girls, unless there is no girl in the family. There are no modern water-works in Persia. The water used for all purposes must be brought by the girls and women from streams. They usually go for water twice a day, once in the evening and once in the morning, to the running streams, which are a short distance from the town.

They never drink the water which is left in the house over one day, but like to drink fresh water every day. In the summer they drink well water. They have both large and small pitchers and jars with ropes fastened to their handles, and after filling them throw them over their backs and march in line homeward. They have several jars made of earthen material, each holding from six to ten gallons of water. One can guess how strong they must be in order to throw such heavy jars on their shoulders without dropping any water or breaking the jars.

It is an interesting occasion when a band of beautiful girls go to bathe in the evening and morning, marching with the jars on their backs to the springs, having no other subject to talk about but their sweethearts. It reminds one of the romantic story of Abraham's servant and Rebecca at the well (Gen. xxiv.), especially of the meeting of our Savior and the Samaritan woman at the well of Jacob (John iv.), because our people practice exactly the same things at the present time as in ancient days.

Nestorian Girls Carrying Water.

PREPARATION OF FUEL.

There is no coal in Persia. They burn wood, principally, which they must plant every year, because it becomes very scarce and costly. The poor classes burn in the oven a mixture of manure and straw. The girls and women during the spring, summer and autumn seasons go about the fields and streets, picking up anything they can find, filling their baskets and bringing them home for the winter's use. They will every morning clean up their own stables and mix the manure, straw and other trash with their hands, and make it flat and thin, leaving it then to dry. Then they heap their fuel in piles for the winter use. They also go out into the forests and vineyards, gathering dry branches of wood and vines, thus preparing everything during the summer for the winter, which is very severe.

ATTENDING TO THE SHEEP.

Some people are herdsmen and shepherds. They live out doors, in the mountains, with their flocks. The women do the milking, the churning, the shearing and preparing wool for the market. They also go into the field, cutting hay or grass, and bringing it home on their backs for the use of the cattle.

THE CHURNING.

This is done in a large earthen jar, which they lay on a wooden saddle. It is a hollow bridge, into which the jar fits; then they pull it forward and backward for an hour before they can make any butter. In order to do this, they first boil the fresh milk, and add some buttermilk to it, leaving it to cool off, and then put it in the jar and churn it. Such is the process of making butter in Persia.

SEWING AND MANUFACTURING.

There is hardly any modern machine work in Persia. Most goods of all kinds are manufactured at home by women. They go to the fields, pick the cotton and bring it home, washing, carding, spinning and weaving it for all purposes. The same with the wool also. Such work is done by women, whether it be for clothing, rugs, carpets, sacks or tents. The Persian women are indeed diligent and virtuous in many respects (Prov. xxxi.).

The sewing is all done by women; the

NESTORIAN WOMAN CHURNING.

sewing machine is not yet domesticated. The garments of both sexes are actually made by hand at home.   There are a few tailors in cities, who make the clothing of the wealthy people, but in the villages, where the masses of the people are usu· ally poor, the women must do all the work.   It will take them several days be- fore they can complete one coat or shirt. Oh, how they need the conveniences of civilization to ameliorate their condition!

THE VINEYARD WORK.

This work begins early in the spring and lasts about two months.   It is the digging and clearing up the space that is occupied by the vines.   The men are dig- ging, the women are piling up the roots of dry grass into piles.   When the men work they usually take off their heavy clothes and put light ones on.   The women at noon and evening must bring their clothes and hats to them, so that they may put them on.   Not only that, but help in bringing them water every hour during the day, and prepare meals for them three times a day.

At the time of grapes the women must pick up the grapes, and in large baskets

carry them to a certain end of the vine-
yard, and put the bunches on the ground
so that they may dry into raisins. They
gather up the dry raisins in piles and
carry them home on their backs for sell-
ing.

THE HARVESTING.

This is not women's work, but in later
years it has become, under the poor circumstances, an occupation of women. It
is very hard work. It begins in the hot
season, and it is miserable work for
women. They have no machine of any
kind. The poor women must take their
little ones with them into the field, a distance of several miles sometimes, having
their heavy hook in their hands, and bowing themselves to the work about ten
hours a day before the burning rays of
the summer's sun. Some women are
compelled to take their two or three
months' old babe in a cradle with them,
and leave it either under the shade of
some tree or sheaves, while they go to
work afar off, not knowing the condition
of the little creature most of the day.

Yet, under all these circumstances, they
seem to be contented, singing all day long
some love songs. It is quite a surprise

NESTORIAN GIRLS REAPING.

to our missionaries when they pass by the harvest fields and hear the women singing under such conditions, thinking they are rather wailing and weeping in such peculiar Oriental tune.

While the women are reaping the men gather up the bundles, bind them into sheaves, and carry them in wagons to the threshing floor, which is quite a distance from the field. The poor women will follow the reapers and gather whatever may fall behind for their support, as it was done in the olden times. Such is the condition of the poor Persian women in the harvest fields. What else could be expected of a nation which claims the inferiority of women?

CHAPTER X.

The Social Life of Women.

THE Persians, as a people, are very sociable, hospitable, and entertaining. They appreciate humor and wit, as well as amusing stories. In this respect they are among the Orientals what the Irish are among Americans. Their sociability is simply a process of their domestic life, but separately classified. The men have their social parties and the women likewise have theirs. The sexes never intermingle unless there is a marriage. Appropriate rules for etiquette, complimentary gifts, with polite conduct in the presence of guests are strictly observed by all classes. But the women lack the advantages of modern society in social life. They have no balls or parties, and no such kind of entertainment as the women of America, with the exception of a few social gatherings each year.

.If they are connected in any way with the wedding couple they will be cordially invited to participate in the festivities of marriage for several days. They must join the rest in the dancing and playing. Then, of course, it is their duty to make the best of the occasion. In such special entertainments the invitations are not only extended to certain women of the family, but to all alike, and they must all go and amuse themselves if they wish to do so, whether old or young; but, as a rule, the young are the participators.

Other social occasions are the national and religious feasts. During these feasts everybody has a right to enjoy the festivities. One of the most enjoyable feasts is that of "Naruz"—New Year's Day. It corresponds to our Christmas in the character of its celebration. It begins on March 14th and lasts three days. It is one of the happiest days in the history of the Moslems' holidays. They begin its celebration on Wednesday and Friday nights of every week, about a month ahead of its real date, by fire-works on the house-tops until midnight. At that time the fields have put on their green garments of herbs and thus say: "Be thou merry in the place of thy youth."

This feast is very Oriental. It was first observed by Zaroostrious, who was much interested in the study of astronomy, and taught that the world completes its movement around its orbit on that day.

PROVISION FOR THE OCCASION.

About a week before, every family makes appropriate preparations for it. The shops and stores all close up, and the bazaars are decorated. The members of the families are well dressed. Eggs are dyed, as well as their hands, feet and hair, which indicates their appreciation and happiness in passing from the old year to the new. The family have in readiness the "Yedi-Lawoon," a composition of seven different eating materials. This is one of the most elaborate preparations of the feast. In the morning of the first day of the feast the servants receive their gifts and are given a vacation for three days of the feast to visit their homes.

The Dervishes, with their hammers on their shoulders, stand at the doorways of the prominent men of the city. There they will stay until they are feasted and given some gifts besides in money or clothing, whatever it may be. If people refuse to give anything, the Dervishes are religiously authorized to demand their due by force.

Eating, drinking and smoking are the order of the day. They make calls, either in the morning or in the afternoon. The calls are usually in groups. They first visit the bereaved families, whose hearts are broken by the loss of their members. Here the visit is very short and impressive in exchanging their sympathies with each other. There is nothing served except the water-pipe smoke.

But in visiting other places they have a good time. When they enter the house, or meet on the streets, they greet each other by shaking hands, "Sallam lekam," "Peace come upon you;" "Biramen Nabarak Ossoon," "Your New Year be a blessed one." These expressions correspond to our "Happy New Year." After exchanging salutations in this respectful manner they are given seats and served with "Yedi-Lawoon," smoking and refreshments. The visitors, of course, do not try to eat all they want at one place, because they will have to visit many other places where they must eat also, but will try to eat very little at each place so that they may keep going until they are through their round of visits.

The people in Persia have no charitable

institutions of any kind like those in America, and the poor have no way of receiving any aid from the people, except on such holidays. In fact, the Mohammedans do not like to help the poor on occasions when nobody will see or know what is done, but they like to help the poor on public occasions, so that everybody can see them doing some good work and glorify them (Matt. 6: 2).

There are four special things which are generally provided for this great feast for the entertainment of friends and relatives. The first thing is "Kylan," a smoking pipe. It consists of a bottle or glass of earthy material filled with water to the top. Through the narrow neck of this bottle passes a wooden tube extending half way down into the water. The upper end of this tube is attached to a vessel of brass, silver or gold, called the "Head of Kylan." It contains the washed tobacco with pieces of burning charcoal on the top of it. Just a little above the neck of the bottle is attached another tube through which the smoke comes to the mouth of the smoker, called "mouth-piece." The tobacco used for this kind of smoke is manufactured at Shiraz, Persia, the best tobacco center in the world.   Both sexes are very fond of this kind of smoke, because it is pure, and in some respects healthy. They smoke about four or five times during the day.   Do not have a pipe for each, but the same pipe is passed around and accommodates several persons at each time.   One smokes about two minutes and then passes it to the next; never keeps it too long, because it is a disgrace to the smoker if he does.   This custom makes the conversation very pleasant and impressive.

Another thing for the occasion is a delicious variety of drinks, such as sherbets, made of plums, cherries, roses, lemons and grapes.   They also drink tea and coffee.   Tea is prepared in the "Samavor," a tea case, and served in small glasses with sugar, but without milk.   Coffee is served in the same way, except the cup is much smaller, but without sugar and cream.   It is as bitter to the taste as wormwood.

"Yedi-Lawoon" is another thing for the time.   A composition of several kinds of fruit, such as candies which are made of the juice of a tree, very pleasant to the taste.   "Pashmak," made of sugar and butter. Sweetmeats, made of pomegranate jelly, which is very delicious indeed. They also have nuts, raisins, figs, grapes and apples, as Persia is a garden of all kinds of fruits.

The last and principal thing provided for the occasion is "Pellow," a preparation of rice in the Persian style. It is first washed and boiled in water with salt. When it is soft and cooked, they drain off the water and let the rice cool. Then they mix with it boiled butter and cover it so that the air may not destroy its taste, and allow the vessel which contains it to stay on the red-hot fire-place for a while, in order that it may be well-done. It is one of the most delightful edibles. They use no forks and knives in eating it, but grasp it by the fingers in handfuls. This is the royal food of the Orient.

The intermingling of Christians and Mohammedans with each other on these occasions must be noted. In many localities they both live together in the same village or town, and it is the duty of each to visit the other during the holidays. When the Christians visit the Mohammedans, they usually take with them some presents; such as eggs, chickens, or a large piece of sugar, as a token of respect. It is pleasant to see them when the Christians visit them with such presents. But Mohammedans do not make friends of Christian people, except to get something out of them.

Men will visit men with such gifts, and women, women. When Christians enter a Moslem's house on such an occasion, they are welcomed even to the extreme of a social meal. They are served with "Yedi-awoon," "Pellow" and delicious drinks. The Christians eat all they can, and take with them what is left for their children, because the Mohammedans will have no use for it. Religiously, Mohammedans are forbidden to touch or eat anything that has been touched by Christians. After the meal is over they must wash the dishes seven times before they are clean, according to the teaching of the Koran. But when Mohammedans visit Christians on "Xmas" and "New Year" day, they bring no presents at all; they just come to eat. Of course, they never eat the Christians' bread and meat, but they eat eggs, raisins, melons, grapes, nuts, and smoke the water-pipe. When they smoke the Christian water-pipe they

put their handkerchief over the mouth-
piece of the pipe, because having been
used by Christians they think that it is
not clean. They are very particular about
physical purity. Whenever they touch
a horse, dog, cat, or any other animal,
the flesh of which is not eatable, they
must wash their hands often before they
are satisfied that they are clean. One
day my mother, who is a Christian, was
playing with the little dog we had, and a
Mohammedan priest saw her touching the
dog with her hands. He said: ''Oh, you
are a bad woman.'' Of course, my mother
said nothing but smiled at him, thinking
his remark foolish.

I then said: ''I suppose if you had
given him some eggs or chickens without
washing your hands, he would not have
refused them.'' Such is friendship and
social visiting between Moslems and
Christians.

Another occasion for social meeting is
in public bath-rooms, what we call the
''Turkish bath.'' They have no modern
conveniences in their own houses for this
purpose; but they have public baths in
the cities, so that everybody may take a
bath for ten or twenty cents. They usu-

ally spend three or four hours in the bath-room, dyeing their hair and feet. The baths are separate for men and women. The owners usually employ female barbers in women's departments and male barbers in men's departments. Persia leads every other country in public baths, which are beautifully constructed and furnished with all necessary things. They have no preparation of quinine to cure a cold in one day, as some of the American medicine companies advertise to do; but any kind of a cold may be cured in a day in a Persian bath-room.

In the villages there are no public baths, but they build a little bath-room in the home and heat the water in large jars, with which they bathe once a week.

There is visiting also apart from the holidays. This is also classified. Each person associates himself with his equals, and visits the same, except in case of sickness or death. Among the highest class, the women send a message to the hostess and notify her of the visit. If it is acceptable, then the visit is made; if not, it is postponed. In visiting, the ladies start with a host of male and female servants, who go before and behind, guarding them on the way, and also as a sign of exalted rank. Upon approaching the place, a servant is sent forward to announce their coming, so that they may be met by the hostess and the reception committee. They are given seats in the reception apartment according to their rank and dignity. There are what they call the "upper" and the "lower" seats. According to the custom, no one has any right to occupy a seat of which she is not worthy (Matt. 22: 12).

After the guests have made their polite bows and exchanged "Salams" (salutations), they take their seats, and are served with refreshments of all kinds. Every twenty minutes the smoking pipe is tendered, and before leaving, coffee or tea is served. This is the mode of living in cities. But when a woman of rank visits the villages, where the people are usually poor, she first sends her servant to the house of the chieftain of the village to inform him of her approach, so that the house may be well carpeted with the best rugs; and if he has not any of his own, he must borrow them from his neighbors for the occasion (Matt. 10: 11; 20: 6).

Upon her arrival, she will be escorted

by her male servants and citizens to her
prepared apartment. If the distance is
several miles, the party will go on horseback; but if it is quite far, she will go in
a "Kajawa," while her servants ride
horseback before and behind her. This
"Kajawa" is something like a carriage.
Two mules are hitched to it. One is behind it, the other in front of it; so that
when one mule goes before, the other follows behind, and they carry it between
them. It is a unique way to ride. It
accommodates from two to four persons,
and has two entrances, one on each side.
No other woman can ride with her who is
not her equal in dignity and rank. As
she is passing by, the citizens will honor
her by killing a sheep or an ox before her
Majesty, and bowing their heads to her.
The head man of her staff will address
the citizens in her behalf, giving them
words of encouragement and sympathy.
She also recognizes their reverence and
respect by bowing to them with a pleasant
smile.

The following are some of the features
of entertainment which the citizens must
provide: Music, playing the guitar and
"Samtar" (an autoharp), which is fol-

lowed by singing and dancing. They usually employ young boys and girls who are professional dancers and jesters, in their outdoor costumes. The Persian women are among the best dancers of the world.

The guests will spend several days in this way. Eating and drinking must be furnished by the poor citizens, which they may not be able to pay for in a month. This is absolutely involuntary on the part of the citizens; but they cannot help the imposition of her Majesty, because she owns them. They are her sheep, and she wants the wool.

CHAPTER XI.

THE MUTUAL RELATIONS OF WIFE AND HUSBAND.

IF THE plan of marriage was carried out according to the law of God, there would be no divorce and no separation of husband and wife in society. The fact is that husbands and wives too often violate this law. And so the further they advance in married life, the darker is the horizon of their happiness. Instead of entering the gate of matrimonial paradise through flowers and lilies, which God opened to them when he said: "It is not good for man to be alone," "What God has joined together, let no man put asunder," to those believing in polygamy, marriage becomes a path of thorns and thistles, a prison of woes and miseries. The ideal relation between husband and wife is truly stated in the law of the New Testament: "Husbands, love your wives, even as Christ also loved the Church." But the unhappy Moslems never apply this divine idea to marriage, because they are ignorant of it; but what they know and say practically is: "Husbands, love your wives when they are beautiful, healthy and wealthy."

Men have, in fact, destroyed the meaning of the word LOVE. They have made it mere human passion. This is because they are ignorant of its origin and purpose. 'Christianity alone gives the real definition of love. It not only teaches and enforces the duty of the husbands to care for their wives, but also urges the duty of wives to love their husbands: "Wives, submit yourselves unto your own husbands, as unto the Lord."

The husbands and wives who live in the sphere of this ideal matrimony will not fail to secure the benediction of God upon them and their children. "Whoso findeth a wife findeth a good thing, and obtaineth favor of the Lord (Prov. 18: 22). Here we see that a wife is the gift of God to man. She is the minister of his comfort, her voice his sweet music, her smiles his happiness, her kiss the guardian of his confidence, her arms the refuge from his troubles and the balm of his health, her industry his common wealth, her economy his safest treasure, her intelligent advice his honest counselor, her bosom the softest pillow of his care, and her daily prayers his ablest advocate with God.

Having considered the mutual relations of wife and husband under the light of the New Testamant, let us now consider these relations under the darkness of the Koran. Under the teaching of the Koran the duties and relations of a husband toward his wife are not comparable with those of Christians, because he does not take her under such promise of united companionship and equality, but for purposes of passion and slavery. He does whatever he may please with her, and if she refuses to submit to his will, she is punished with death. In case of sickness and death he renders her no help or sympathy. One may ask: "Who is to be blamed for such disorderly and disgraceful conduct, he or she?" I reply: "Neither of them, but the teaching of the Koran." It commands thus: "Those wives whose perverseness ye fear, remove them into the harem and beat them to your satisfaction; but if they submit to your will, do not bring any charge against them without a cause. If any of your wives be guilty, select four witnesses from

among yourselves against them, and if
they testify against them, sentence them
to imprisonment; let them die there."

The women taken in the act of sin are
put in woolen bags, the openings of the
bags tied, and the women beaten with the
hands and kicked with the feet as if they
were foot-balls and clubs in a gymna-
sium, because they have brought public
disgrace upon their husbands. Even
their own relatives do not sympathize with
them. Such is the law and practice of
Islam.

Another mode of punishment is to put
them in a deep cave and bury them up to
the shoulders, while standing on their
feet, and plaster them in the ground so
that they cannot move till they are dead.
These punishments are only for bad
women; but good women will be punished
at home in some other way. If a man
does not beat his wife he is thought to be
a crank. Sometimes he beats his good
wife only for the sake of public opinion.
Mohammedan women often come to Christian women and ask them to send missionaries to preach to their cruel husbands, and induce them to be kind; because there is not a single particle of

advice in the Koran that might lead the husbands to love their wives. More than this, when the husband eats the wife cannot eat with him; she must wait on him first, and eat afterwards. When he is through eating she must bring water and wash his hands and wipe them with the towel, and prepare his water-pipe, so that he may comfortably smoke. Then she goes into another part of the room and eats with the children and other women, in order that he may not see her moving her lips and finishing the meal. When he is walking in the street she must not go near him, because it is a disgrace to him when the people see her with him. Women and their husbands never walk side by side, as they do in America, but the husband goes first while the wife follows him, or else they must go by themselves; hence they have no public pleasures with their husbands at all. Often the husband goes somewhere without saying good-bye to his wife, and he may go for many years without writing any letters to her. Anything needed to be done in the home he will write to his male relations about, without saying a word to his wives. Moreover, when he is transacting business, or is in trouble of any kind, he never mentions the fact to his wives or asks their advice in the matter. If the wives have any mental or physical trouble they do not dare to let him know, because they are pretty sure he will, instead of sympathizing with them, give them a good shaking. Among the high classes the wives receive better treatment at the hands of their husbands than in the lower classes. They always call their husband "lord" in addressing him. The expression "papa," or "dear," is not in use among the Mohammedans, because "papa" and "dear" signify an equality in the two persons speaking; while "lord" signifies the superiority of the one over the other.

Women who are mistreated in such ways make every possible effort to repay their so-called husbands for their unjust deeds. Owing to these exciting conditions in the family, the home becomes a den of wild beasts. The children inherit the same troubles. The love of father, and mother has no place in their hearts. While this is the condition in general, we are glad that there are some men who are naturally kind-hearted, even among the heathens. Some husbands treat their wives justly, and some wives, too, are naturally endowed with talents and kindness, which they use in elevating their husbands to a higher culture. But, generally speaking, the men and women of natural talents and common sense are very few, and the others must be taught and developed to a higher standard of life.

CHAPTER XII.

WOMEN IN SICKNESS AND DEATH.

AT THIS critical condition of life, the Persian as well as all the Oriental women are in a sad state. The severe attacks of sickness on the one hand, and the carelessness of the unsympathetic husband on the other, often result in death. There is no physician to prescribe for them in all the country, with the exception of a few missionaries who are stationed in the principal cities. The women of the villages are sick for weeks and months and years without any medical help, and finally die.

Regarding the immortality of women, the Islamic teaching affirms that women have no souls; that they are objects of pleasure so long as they live, and when they die they are no longer in existence. This is the popular belief of eighty-five per cent. of Mohammedans in Persia. Three facts account for this: first, because the Koran claims the inferiority of the women; secondly, paradise, or heaven, has seven departments, as follows: the garden of eternity, the mansion of peace, the mansion of rest, the garden of Eden, the garden of pleasure, the garden of meeting, and the garden of paradise. All male believers will be led by the angels into these departments after death, and will be given other wives much prettier than those of this world. These wives of paradise are called "Hoory" and "Pary," the most beautiful gifts of God. The men will also be given anything that is best in eating and drinking. They will have no more sorrows, no more sickness and no more death, but together with their new wives will visit the different gardens of paradise and have an everlasting good time. But the earthly wives will perish unless they be subject to their husbands and follow faithfully the dictates of the Koran. They believe such wives, as keep the faith, will go along with them to these paradises, but not the others. The number of new wives in paradise is limited to seventy-two for each believer; and if he perchance has his earthly wives also, that will be far better.

Some wives strive to attend all the religious services, and be subject to their husbands in all things, so that they may follow their husbands to heaven after death, while others make pilgrimages to Mecca, where they may obtain favor from the grove of the Prophet. They make the journey on donkies and mules as a sign of humility, so that their sacred pilgrimage may be better appreciated by the Prophet and he may grant them their wishes. If some of them die before they visit Mecca, their bones are boxed and taken there by some of their relatives and friends, in order to gain for them immortality in paradise. Such is the religious hope of Persian women without faith in Christ as the only life and author of immortality.

In some parts of the Orient sickness is worse than death. When a person is sick, the people think that the devil has gotten in him, and they try to beat him with clubs, so that the devil may go out of him. While in other parts the doctrine of predestination dominates the minds of thinkers, who believe that God has appointed a certain time to live and to die, and therefore it is no use to care for the sick and afflicted. This fanatical idea is very common in the Orient. Often the patient dies without the slightest knowledge of friends and near neighbors. The most common diseases are those of the eyes and throat. The first attacks both sexes and all ages. The inflammation is very severe in the Orient, and more common in females than in males. One visiting the Orient would be surprised to see such a large percentage of blindness among all classes and races. They will treat each other by injecting all kinds of powders into the eyes without any knowledge of the parts. Whenever the patient is injuriously affected by such powders, they inject sugar in order to sweeten the surface of the eyes.

The disease of the throat, or diphtheria, occurs in the forms of tonsilitis, pharyngitis and laryngitis, which usually attacks the children epidemically. I remember my little nephew was subjected to this disease. He could not talk loud enough to be understood, but made grimaces, and begged us for help; but we could render him no more help than he could himself, and he passed away peacefully and helplessly.

Another prolonged disease is that of insanity. A woman who becomes subject to this most distressing and absolutely helpless disease for years is treated worse than a dog by her Moslem husband. As we have already indicated, Mohammedan women are loved and cared for, possibly, when they are beautiful and healthy, but in such a sickness, in its indefinite course, they must simply give up this world at once, because the husband, and even their relatives, will tire of them.

It is no wonder that Christ was so active and compassionate when he was engaged in the divine practice of curing the forms of disease for which the medical profession has, as yet, not found any curable methods. The people of the Orient have not had such benefits bestowed upon them since the time of Christ and his followers. Has he deserted them? Nay, but he said then, and says now, "Go ye into all the world and preach my gospel to every creature and heal all kinds of disease." My Christian friends, remember that Jesus was not only a teacher, a preacher, but a physician also; yes, the greatest physician the world ever saw. "Take they bed and walk," was his most marvelous cure. You cannot deal with men in heathen lands spiritually, unless you first care for them physically.

This is the very key to the door of their hearts. Would to God that we as Christians might master the divine ideas of Jesus! I wish we were willing to learn his method and means of healing, by which he so deeply impressed the minds of poor, suffering humanity. God help us to love and follow Christ, who gave his life to elevate mankind from sickness and sin to everlasting health and happiness!

In case of malarial, scarlet and other kinds of fevers in Persia, they take the patient to the priest, who makes charms and spells in the name of Mohammed, thinking they will be relieved. There is no such thing as a "State Board of Health" to prevent disease and protect the people from deadly superstitions. The priest receives nothing more for his simple and false treatment than a few eggs and chickens.

BURIAL CEREMONIES.

In the religion of Mohammed the ritual of burial is elaborated according to the circumstances of the individual. If a woman is poor, the burial will be short in ceremony and poor in entertainment.

Women usually receive very little respect at the hands of their husbands, even in death.   When the sad news is reported of the death of a woman, the public will hardly utter any sorrowful expressions, unless she leaves several children behind. The announcement of a death is regarded as a calamity, and one seldom volunteers to carry the news to others.

The "Moolloh" (or- herald) must be first notified of the fact, and he will make it known to the public by going to the top of the mosque and by calling to prayer in a loud and wailing voice. They use no bells for that purpose, think- ing there would be too much of music in ringing bells for such an occasion.   In fact, they never like to hear the sound of bells at all.   Whenever the Christians use bells the Mohammedans steal them. After everybody has heard the voice of Moolloh, the preparations for the burial soon begin; because they never keep the dead body in the house for two or three days, as they do in civilized countries, but bury it within four or five hours. They suppose that it will cause some psychological effects upon the family and they may likewise die.   Yielding to such.

superstitions, they often bury the individual when not dead, but in a state of unconsciousness. When one is unconscious for several hours, believing that the person is dead, they take and bury him.

According to custom, they wash the body in a private room for an hour, till it is perfectly clean; then put cotton in the nose, ears and mouth, so that nothing can fall in after burial. The body is then clothed in a white garment, and sometimes placed in an ordinary wooden coffin, which they make at the time, but the majority are buried without it. The reason is, there are no undertakers who provide coffins when needed, but the coffin must be made when the person dies, and often they cannot get any one to make it on so short a notice. There is not a single coffin in all Persia that looks like those we have in America—so beautiful and valuable. In fact, if men should spend money and make such, they would be stolen on the first night after burial.

The washing of the body and digging of the grave are done by poor people, who receive usually some money, together with the clothes of the dead for their services. When the body is ready, they send after Moolloh to come to the house. Then the body is taken to the cemetery upon the shoulders of six men, while the others follow, and when the bearers are tired others take their places till they arrive at the grave. The Moolloh and his party go before the funeral procession, reading some passages from the Koran in a loud voice. Women are not allowed to go to the graveyard, but must stop at the edge of town for a few minutes and then return home.

There are no carriages or professional undertakers in Persia, but everybody will render timely assistance on such occasions. The grave is about five feet deep. The body is interred on its side, with the face looking towards Mecca, the holy place of Moslems. At the center of the grave, right above the coffin, they extend big timbers from one end to the other, over which straw is placed, and above the straw earth, while two young men get in and trample down the earth with their feet. The graves are about two feet higher than the ground.

On the first evening of burial, they build fires over the grave, supposing this will prevent wild beasts from digging into the grave for the dead, as the wolves do in the Orient.

In Persia there are very few marble monuments over the graves. Most of these are common stones whereon the name and the date are written. They have plenty of flowers and roses, but never regard the dead by putting them on the graves. The cemeteries, instead of having fences, gates or trees, are open courtyards for all kinds of animals. After the flesh has left the bones, they open the graves, take up the bones, pack them in boxes and take them to Mecca, where they bury them beside the grave of the prophet, awaiting his resurrection at the last day.

The burial completed, the Moolloh and the people return to the house of the bereaved and pay their condolences; "It is the will of Allah, may his name be blessed." Then they are served with very bitter coffee, as a sign of sorrow and mourning. In addition they have a bottle of rosewater, dropping some on the right hand of each one, and then leave for their own homes.

The period of mourning varies from

three days to a month, according to the influence of and regard for the individual. The usual time for the poor is three days, the last of which is the special one. The Moslem husband does not weep over his lost wife unless he has several children; because he thinks it is not a hard matter to find another one, there is no use for him to weep for her. But when a man dies, the entire community and family mourn for months, and even a year. Among the higher classes the women receive more attention in the burial ceremony. The husband on such an occasion will kill several sheep and feed the people who attend the funeral for three days; on the last day the family and the near relatives are dressed in black garments which they wear for several months.

It is interesting to note the manner in which Persian women bewail the dead, especially when a person is wealthy and handsome. There is not a single hymn sung in all the Orient. But they have professional weepers, who are hired for the occasion and sing death songs. Two or more sit by the bedside of the dead, together with many other women. Then some one hands them the clothes of the dead and they take them in their hands and weep, singing the death songs. While they are singing the other women keep time.   At the last song they all join in the chorus in a very loud voice, then the women will take the lead again, and so on for five or six hours. When the professional wailers get tired they hand the clothes to some other women who have recently lost their dear ones, and then they will take the lead till all are utterly exhausted. However, they will keep on until the body is taken away for burial.   The ideas involved in the songs of these professional wailers are related both to the parents, relatives and the dead person herself, celebrating the special deeds of the dead and the sorrows of the living parents and husband (Matt. 2.: 18).   They sing in such sympathetic tones as to cause everybody to weep, and attract a large crowd.

There is another peculiar thing on this occasion: the mothers, sisters and wives tear their garments, pull their hair, scratch their faces, and plaster their heads with mud and dust; while the brothers, fathers and husbands keep their shirts unbottoned at the breast and beat themselves severely, thus showing their affection for the dead. It must be noticed that they do this when a man dies, but not for a woman.

On each Friday night for a month the Moolloh and some of the relatives will go to the graveyard, where the Moolloh will read passages from the Koran, while the others distribute edibles to the poor, supposing this will benefit the dead. Such is the hope of Moslems in death and eternity. Persia is not the only country where death is an awful gulf of darkness for women and men without Christianity, but in any land where the name of Jesus is not preached and believed in, death brings woe and misery both to the living and the dead. Instead of singing the songs of peace, love, and hope in the supreme name of Christ who rules death and life, they sing the songs of superstition and hopelessness.

CHAPTER XIII.

CHRISTIANITY THE ONLY HOPE.

CHRIST and his apostles have left us sufficient facts with which to contrast Christianity with other religions, and to see the high dignity and honor which the Christian faith confers on woman. In no other religion is woman equally yoked with her husband in all that belongs to man. This regard for woman is one of the strongest evidences of Christianity and of human progress. Among the Greeks and Romans we find women in a state of degradation and inferiority. Men had all the advantages of learning and culture, while women, with the same God-given faculties, were deprived of what really belonged to them and doomed to the extreme of ignorance and to the barbarity of passion.

Mohammedanism claims to be the greatest religious system; but with all of its morality, alleged inspiration and belief in "Allah" as the one and only true God, it has done nothing to elevate woman. It has made her the tool and slave of man.

Turning to Hindooism, the great religion of India, we find it the path of darkness and despair for woman. It takes away what absolutely belongs to the woman. She is considered incapable and unworthy even to render worship to the gods. Her lord is her husband. To him she must offer her sacrifices and her devotion.

Buddhism also means the enslavement of woman. This is the second great religion of the world. But all its gods and goddesses of brass have reflected no light upon the degraded, helpless condition of women. The attitude of Buddhism towards women is even worse than that of Mohammedanism and Hindooism. It teaches that because woman first sinned, she brought the curse of evil upon the human race, and is worthy of all kinds of punishment.

An intelligent Chinaman said to a missionary: "Why do you make Christians of the women?" "To save their souls," said the missionary. "But they have no souls; how are you going to save them?" Such is the idea of the heathen world without Christianity.

But above all, I thank God that Christianity has planted divine truth in the nature of woman as well as in man, and that this truth brings perfect moral and physical health. The very object of Christianity and its self-sacrificing followers is to carry glad tidings of joy to the destitute and broken-hearted, healing all woes and ills. This is Christ's message of equality to all nations and races and both sexes.

By Christianity, I do not mean superstitious forms of it, practiced by some denominations and in some countries, but I mean the pure, sincere, loving and peaceful teaching of Christ as it is declared in the gospel. I have seen in some cities and towns of Europe that woman's condition is not any better than in heathen lands. One often sees in Europe women yoked together with dogs to carts, carrying fruit and vegetables to market, while their husbands drive them, and yet they claim to be Christians. Christianity, applied in daily life, is the very foundation of hope and respect for women.

Christ did not only teach his followers, but he actually lived what he taught. He rendered to woman the love and respect due her. His unselfish zeal and earnestness in giving the gospel to the Samaritan woman at Jacob's well is one of the strongest testimonies of what Christianity means for woman. On another occasion, his divine defense of the woman taken in adultery, before the supreme court of the Pharisees, is another tribute of Christianity, "He that is without sin among you, let him cast the first stone." Yes, he was the best friend of woman in all circumstances. He always took interest in the domestic and sacred life of women. He healed their sick and raised their dead, using all his human and divine powers among men and women in order to bestow upon the human race his spirit of righteousness and love of equality. When he ascended to heaven he committed all such noble and holy examples unto his beloved disciples, and urged them to preach the same blessed gospel of peace and love to all nations.

So to-day Christianity has the same mission—to carry freedom, equality and salvation to the men and women of all races.

As the result of Christ's self-sacrifice and the church's achievement, Christianity is the great promoter of education and redemption, the mental, moral and spiritual beauty of women.

And so Christian women shine like the bright star of the morning everywhere in the world. They have organized their spiritual and intellectual forces, and are fighting the evil of the whole world with clear, conscious might. They fear nothing; Christ is their leader, the gospel their strength, and heaven their eternal reward.

May God help all women to acknowledge their true mission, join this great army of Christian workers, and advance into all heathendom to secure the dignity and rights of their sex.

THE NON-MOHAMMEDANS.

In Persia there are about 30,000 Jews, 60,000, Armenians, 15,000 Fireworshipers and 100,000 Nestorians. Let me first call the attention of the reader to the Nestorians, in which sect I was reared.

Nestorius was born in Antioch, and was a disciple of Theodore of Neopanestru. He became a presbyter of Antioch, and was made Patriarch of Constantinople in 428. He was distinguished as a man of vast learning and zeal of speech. While

in Constantinople he was invited to attend the general assembly of the Roman
Catholic Church, in order to discuss the
divinity of Christ. He opposed the dogma
of the virgin Mary as being "Mother of
God," because he believed that there were
two natures in Christ; that Mary was the
mother of his human nature and not of
his divine nature, and that God is not
limited to space, time, birth or death.
Nestorius was opposed by many of the assembly, but a few followed him. Then
his opponents cursed him, and he was ex-
communicated in 430 by the Council of
Ephesus and Pope Celestine. Nestorius
proceeded to establish his own doctrine.
He had a seminary in Constantinople and
Odessa, and prepared about three hundred missionaries, sending them into
China, Japan and India for the propaga-
tion of the gospel of Christ among the
heathen. There are a good number of
Nestorian Christians now in those
countries, having inherited their faith
from their forefathers. One day a Jap-
anese was asked what was his belief.
He said: "I am a Nestorian." After a
course of consecrated life Nestorius died
in Upper Egypt, but his work is still con-
tinued by his followers.

The present Patriarch of the Nestorian Church is Mar-Shumon, whose residence is in Koorchanis Kurdistan. He is regarded by the Nestorian people in the same manner as the Pope by Roman Catholics, except that he does not claim to forgive sins. Two years ago the Pope sent a message to him and asked him to come back to Catholicism, but he refused. He is the last of one hundred and eight Patriarchs of the Nestorian Church. The name ''Nestorian'' is given by the Pope, but the people who are called by this name are Syrian by race, and were driven to Persia and Kurdistan by the persecutions of the followers of Mohammed. The total number of Nestorians who live in Persia and Kurdistan is 150,000, with Mar-Shumon as their Patriarch.

The Nestorian Church practices immersion as the mode of baptism, but baptize infants. They administer the Lord's Supper in the same manner as Protestants, but with more ceremony. The church does not forbid any one from taking the Lord's Supper. The minister stands before his congregation and states the conditions under which they may take the supper, and then lets them decide for themselves.   Everybody who believes may take the Lord's Supper, whether young or old. They fast twice during the week, on Friday and Wednesday. They also fast fifty days before the resurrection of Christ, and twenty-five days before Christmas. During this time they eat no animal food, such as meat, eggs, butter, but vegetable food, and give themselves to prayer.

No other denomination has suffered so much persecution for the sake of Christianity as the Nestorians. Since the beginning of the Christian era they have been robbed and killed by the heathen for the sake of their faith in Christ. Their seminaries and schools have been destroyed, their libraries have been burned and their churches ruined, but nevertheless they have not been robbed of their everlasting faith in Christ. They prefer to lose their property and lives, and suffer all kinds of torture from the Kurds and Mohammedans, and die in Christ Jesus.

The Nestorian women are much like Mohammedans in manners and customs, because they live together in the same towns and consequently are to a great extent influenced by them in the mode of life. They have various occupations and social and domestie accomplishments. But they differ from Mohammedans—first, in the ceremony of marriage; and secondly, in costume.

Their marriage ceremony and entertainment is about the same as that of Mohammedans, except in the case of the minister. The ceremony proper usually takes place on the last day of the marriage program. It must be celebrated at between four and five o'clock in the morning. The young couple and the minister must not eat anything until the ceremony is over. As to the purpose of fasting, I cannot give any information. The minister conducts the marriage in the presence of several respectable men and women. The bride and groom, with their best friends, stand before the minister to be questioned in the same way as in America. But the bride must be veiled so that no one can see her face, and in answer to the minister's question she never answers "yes" or "no" so that the people hear her, but if she wants to say "yes," she bows her head, which indicates her wish; and if she wants to say "no," she inclines her head backwards, which indicates her refusal.

The groom usually answers the minister loud enough for everybody to hear him. After the minister is through the questions, he takes a cup of water. and wine, and says, "I mix the water with the wine; whenever they separate from each other, then you may have the right to leave each other." He gives the mixed drink first to the bride. She drinks half of it, and then hands it to her husband, and he drinks the rest.

This means that symbolically they can never leave each other, so far as they are consistent in their lives, because it is impossible to separate the wine from the water.

The costume of Christian women is entirely different from that of Mohammedan women. It is difficult to give a description of it, and I refer readers to the illustrations in this book. Polygamy is not practiced among the Nestorians. The women are loyal to their husbands and faithful in the house, but they lack the advantages of education. They work very hard in helping their husbands.

Women whose financial condition is

good, dress very much like American women, with the exception of the head-dress (veil). Their occupation is embroidery and supervision of domestic life.

The little light that is reflected from the life of Nestorian women, as well as of men, is due to the good work done by Presbyterian missionaries. When the missionaries came to Persia they found the Nestorians in a low condition. They had lost their education and, to some extent, their religious principles, because of the severe persecutions which they had suffered at the hands of the Kurds and Persian Mohammedans. Their ancient Syriac language was transformed into the modern Syriac, which is the dialect used by all Nestorians in Persia and in Kurdistan. But their religious books were still in Ancient Syriac. No one but the ministers were able to read them. There were only four Bibles in all Persia, one of which now belongs to my cousin, left to him by his ancestors, who were ministers. It contains only the four Gospels, and it is about fifteen hundred years old. This is the only Bible in Persia coming down from that ancient time. The other three were sold and taken to the British Mu-

seum.   On several occasions the missionaries offered one hundred dollars for it,
but my cousin, who is a Nestorian minister, would not sell it.   Women as well as
the men reverenced this copy of the Bible
very much.   People come from all parts
of the country, bow down, and kiss the
Bible three times, but they do not understand what is in it, because they cannot
read it.   My cousin has wrapped it in
layers of fine cloth, keeps it in a trunk,
and never uses it in the church service.

There is so much difference between
the Ancient and Modern Syriac that when
the minister reads the Ancient, the people who talk the Modern Syriac cannot
understand him; consequently, the people do not get any benefit from it.   But
since the missionaries came, the people
have been supplied with Bibles in Modern Syriac and are taught in complete
editions of the Bible, instead of four
books.   Now, as a result of the mission,
we have a great number of our women
and men both morally and intellectually
prepared for their life-work.   May God
send more of such devoted reapers to his
harvest fields.

THE ARMENIANS.

Americans have read a great deal about the Armenians as being a Christian people. They number about 60,000 in Persia, and those who live in cities are of good influence and means. The nature of their business is such as to bring them in touch with the government. They control some of the best stores, shops and offices, just as they do in Turkey. But those who live in towns and villages are in the same condition as the Nestorians. Their manners and customs are the same. Their women are of the same disposition and are as ignorant as the rest, except those who are wealthy.

Their church government is like that of Roman Catholics. They practice infant baptism by sprinkling, claim the Virgin Mary as the "Mother of God," and have private confession before the priest. They are subject to a good many fasts and superstitions, but their faith in Christ is very strong. If necessary, they will give their lives for the sake of Christianity.

One thing is very much needed among the Armenians and the Nestorians, and that is a new birth in Christ. "Ye must be

born again;'' and when they are, they
will be among the best Christians in the
world.   The doctrine of regeneration has
never been understood by them, because
they have not been taught, except in the
case of those who have missionaries. Oh,
how they need the full spirit of the gospel!

THE JEWS.

The Jews, who number about 30,000, are
absolutely without the gospel. They are
as blind spiritually as they were in the beginning of the nineteenth century. They
generally speak Hebrew among themselves, but they can talk the Persian, Armenian and the Nestorian languages, with
which they come in contact in business
transactions. They are money-makers, as
the jews are everywhere. Their women are
in a bettter moral condition than those of
other heathen countries. They dress and
live in the best style, but suffer very much
at the hands of the Moslems. Those who
live in cities can hardly go out-of-doors,
except among their own community.

THE DEVIL WORSHIPERS.

These people dwell in the mountains
in Central Kurdistan.   They believe that

the world belongs to the Devil. He is the prince of it—to him is due reverence and worship. They admit that the Devil was cast down from heaven, but they believe that in a certain time God will forgive him and restore him to his angelic office. They think of the Devil as we think of Christ for our future hope. Their priests only take care of the golden image of a rooster and keep it in one place. Often people will offer a large sum of money to the priest and take the image to their homes for a certain time and worship it. On certain occasions they have public worship, taking the image and placing it on an altar; then the priest pronounces his benediction upon it. After this a napkin is spread at the feet of the image and every one comes forward and puts his offering on the cloth. The money is given to the priest. They don't spit on the ground, because they believe it is the devil's face. If one dares to say anything against the Devil, he has a hard time to away safely. They will not pour hot get water on the ground, fearing to scald the Devil.

Their social customs are altogether different from the rest of the Persians.

Women and men on certain occasions freely associate with each other in dancing or marrying, whether they are sisters, mothers or brothers. Parents, as a rule, sell their daughters for a sum of money, just as they do land or cattle. This, of course, is not compulsory on the part of the girl; if she does not prefer to get married, she can remain at home.

One thing is very impressive among these heathens: they believe in their religion with all their mind and heart and money, whether it does amount to anything or not. Here is a great object-lesson for Christians in America.

An Appeal to Baptists.

❧ ❧ ❧

IN THE foregoing chapters I have described in a simple way the condition of the women of Persia, who are without the gospel of Christ. Let me now present a few thoughts concerning the grave responsibility resting upon Christians, and especially upon Baptists, who claim in church government and ordinances to conform to the Apostolic Church.

Can we be called faithful and loyal disciples of Jesus Christ unless we obey his command in giving the gospel to all nations as the sovereign cure of all the woes of humanity? Surely the church of Christ cannot attain the noble, sublime perfection of her Lord and Master until she honestly seeks to evangelize all nations. "As the Father hath sent me," said the Christ, "even so send I you." He was God's first great foreign missionary. And if the Christ-life is in us, it must impel us also to be missionaries of the cross "to every creature."

Many Baptist Churches are missionary

in spirit and endeavor. The work of Baptists in foreign lands during the last century is the brightest page in the history of our denominational life and achievement. In China, India, Japan, Africa and many other lands our mission stations are beacon-lights of truth.

But have we no duty to the land of the Bible, the country in which the Christ himself was born and lived and died for us? Ought we to cease our missionary efforts when, as Baptists, we have not a single missionary in Persia or in Palestine, where we claim the first Baptist Church in the history of the world was established? The people of the Orient have known nothing of pure Christianity and its blessings since the apostolic age, and it would seem that we owe them a debt which can be paid only in the coin of gospel grace. As descendants of the original mother-church, let us meet the ancestral obligation of gratitude and love by sending the dwellers in the far East the gospel and its ordinances in their primitive form!

Would not the exalted Christ, in the place of power above, smile upon such a missionary undertaking, and feel a sense of satisfaction in beholding the standard of the Cross planted anew in the land of his human birth and divine sacrifice for men? When on earth he felt the bonds of obligation binding him to the altar of sacrifice, not the least of which were the cords of love and the bands of a man. And any earnest effort on the part of his followers to-day to redeem the Holy Land from the hands of the infidels, must fill him with unutterable joy, for then he would see of the travail of his soul and be satisfied.

Persia, and indeed the whole Orient, presents an open door and blessed results to gospel missions. The Christ seems to say: "Behold, I have set before you an open door, and no man can shut it." Quickened by his spirit and grace, let us enter in and possess the land for Christ!

The claims of suffering humanity demand it. The woes of womankind require it. The debt we, as redeemed sinners, owe to Christ, who died for us, renders this forward movement of the Church imperative.

Let us, then, be up and doing! Let us carry the "good news" of salvation to the lands where Christianity won its first victories.

www.ingramcontent.com/pod-product-compliance
Lightning Source LLC
Chambersburg PA
CBHW080246180526

45167CB00006B/2431